Self-Made Man

East African Mammals (7 volumes)

Island Africa

African Mammal Drawings (facsimiles)

Kilimanjaro: Animals in a Landscape

Arabian Mammals: A Natural History

Self-Made Man

HUMAN EVOLUTION FROM EDEN TO EXTINCTION?

JONATHAN KINGDON

John Wiley & Sons, Inc.

NEW YORK · CHICHESTER · BRISBANE ·
TORONTO · SINGAPORE

Library of Congress Cataloging-in-Publication Data

Kingdon, Jonathan.
    Self-made man : human evolution from Eden to extinction? / Jonathan Kingdon.
      p.   cm.
    Includes bibliographical references and index.
    ISBN 0-471-30538-3 (cloth)
    1. Human evolution.   2. Human ecology.   3. Man, Prehistoric.
  I. Title.
GN281.K49   1993
573.2--dc20

To the unknown child
Everyone's past, humanity's future

Contents

Preface and Acknowledgements

This book began with the simplest of ideas. All animals adapt to circumstances and circumstances change. A species is the sum of such adaptations.

For humans, by definition tool-making animals, 'circumstances' have become more and more self-made. I reasoned that physical, psychological and perhaps social adaptations must have developed in response to challenges posed by technology itself or to an environment mediated by tools and artefacts. Furthermore, as humans spread out far from their native African savannas they put themselves into very different self-made predicaments, each of which demanded its own adaptive response. For example, an animal that becomes stranded eventually changes and develops into a unique island-dwelling form of life. If the agency for that stranding is a self-made artefact then technology acquires a special evolutionary dimension. Thus, Aborigines can be said to owe their biological distinctiveness to a single invention – the watercraft that landed them in Australia. Exploring whether the development and diversity of Modern humans might owe as much to technology as to nature is one of the strands of thought in the pages that follow.

A more personal motive came from that almost universal desire to learn about the family to which I belong. The sources I turned to for an authoritative account of human origins resembled disjointed parish registers full of torn pages. Amongst the mass of conflicting ideas and theories were plenty of interesting facts but few attempts at comprehensible explanations for the dispersal and radiation of Modern humans. I knew the diversity of my brothers and sisters must have a biological explanation. I knew that my family had dispersed all over the world. Such certainties drop away fast when the present is extended back into the seemingly unknowable past. Yet this unknown is punctuated with solid fragments of certainty. Fossils and artefacts scattered through time and space mark out a pattern. Occasionally our probing fingers can entwine with the digits of our dead ancestors. We can work backwards or project development forward from simpler beginnings.

Today patterns emerge from many more fragments of knowledge than bones, and the central authority of the classifiers and categorisers has given way to more diversified attempts at recreating events. Our ancestors, particularly the more recent ones, lived lives what were at least comparable with our own in the vivacity, if not complexity, of their social intercourse. Their relationships with other animals and plants were generally of far greater intricacy than our own.

Ancestors conjured up by contemplation of the shop-soiled world into which we have been born are different from those reconstructed in a museum or textbook. In my own case, curiosity about origins did not begin in a classroom, laboratory or library. It began on my own home ground in Africa. The landscapes of my childhood still rumbled with the passage of extinct herds and I could relish the same lake breezes that had cooled thirsty hunters a million years ago. These were not waters to be dredged for scientific fodder or fields to be ploughed for data, but a land where I was only the most recent adolescent inhabitant.

An awareness of prehistory, especially for an over-energetic youth, needs to be awakened by exceptional experiences and reinforced by the enthusiasm of sympathetic mentors. For me there were excursions to Olduvai Gorge and expeditions with the Hadza, the last hunter-gatherers of Tanzania, or fishermen at Mjemwema and Kunduchi augmented with visits to the Coryndon Museum Hall of Man and to Phyllis Ginner's verandah in Dar-es-Salaam to pore over sheafs of tracings from the rock art of Kondoa. Later I would camp at rock art sites on three continents and see many prehistoric paintings at first hand.

In the slow growth of experience and ideas about early humans I am deeply indebted to many friends and colleagues, including all those who only reached me through books, papers and letters. However, memories and affections fuse where days and nights have been shared under an open sky in wild places. There were early campings with parents and family friends in many parts of Tanzania. More purposeful and inspiring was my first dig at Napak, in Uganda, with Bill Bishop and Sonia Cole. It was Bill Bishop and Alan Walker who first impressed upon me that human prehistory was not the closed preserve of an elite cognoscenti but an open-ended field in which insights could be found, and were sought, from a much wider scientific field. Here my own interest in mammalian ecology and evolution seemed to mesh in seamlessly. At about this time Julian Huxley had pushed me towards writing my first book in which I portrayed *Homo sapiens* as one species of the many that inhabit East Africa; he encouraged a vision of plants and animals enmeshed with the land and its seasons, all existing in a continuum that stretches way back to long before there were humans.

Alan Walker, Colin Groves, Clifford Jolly and John Harris have maintained a sympathetic and stimulating fellowship over many years. Colin Groves, Alison Jolly, François Bourlière and Elspeth Huxley have read the manuscript and made many helpful suggestions. As a research fellow of the Kenya National Museum I had the time and the opportunity to see the huge numbers of fossil hominids and to discuss and contemplate the possible meanings implied by all that diversity of form. On a trip to Koobi Fora with Richard Leakey the pursuit of human origins acquired a fascination and grandeur that still seems as vivid as my first impression of the bleached shores of Lake Turkana and the fractured lava fields of the Suguta Valley.

No naturalist brought up in East Africa can fail to acknowledge their indebtedness to the Leakey family for all they have done to bring life and vitality to the search for human origins in this hominid Eden. Their influence has ranged from the personal and particular to the public and institutional. They have nurtured an intellectual climate which was certainly crucial for the growth of my own ideas; so to L.S.B., Mary, Richard and Meave a heap of thanks.

The ideas expressed here have had a long time to mature and this is very much a 'second generation' book. The early fossil discoveries were surrounded by heady, breathless excitement and feverish controversies. Neanderthals, 'Pekin Man' and the Southern Apes, all former stars in the firmament of human origins, have settled down into less central positions within a broader evolutionary pattern. Personalities and controversies are still very much alive and the players figure prominently in most books on origins. In the pages that follow I have deliberately tried to limit the intrusion of contemporary scientists into an essay that focuses on the lives of prehistoric people. This may divest some original ideas of their authorship and blur distinctions between the more and less probable, but it has been the only way to get an unashamedly hypothetical narrative less cluttered with modern busybodies. We all know that a huge rootstock lies hidden in the soil of prehistory and I am aware of the dangers of deriving concepts from today's appearances and trying to graft them on to such hidden prehistoric roots.

If frequent quotes and citations help insure authors from responsibility for controversial ideas I should at least give credit for one unauthored strand of argument that is central to this book. It originates in part from some old ideas that, for obvious reasons, never received popular acceptance. Who would believe that Africa's most enduring symbol, the darkness of its contemporary population, may actually be the product of ancient incursions out of Asia? In 1919 A.C. Haddon (an influential mentor in the careers of L.S.B. Leakey and Solly Zuckerman) suggested just that. He in turn was influenced by ideas of a prehistoric Afro–Asiatic connection initiated by J.L.A. de Briau and E.T.J. Hamy in Paris at about the turn of the century. Only the most recent discoveries in genetics, archaeology and physical anthropology have made that outlandish conclusion not just plausible but very likely.

Most of the authors who have influenced my thinking are listed in the bibliography. To these must be added the many friends and colleagues who have contributed in more ways than can be detailed here. They are (in alphabetical order): P. Agland, M. Aitken, P. Andrews, M. Archer, C. Bangham, K. Behrensmeyer, P. Bellwood, F. Bourlière, C.K. Brain, G. Brauer, B. Breeden. T. & J. Butynski, J. Callaby, G. Caughley, M. Coe, J. Chappel, S. Cole, P. Collett, H.S.B. Cooke, Y. Coppens, F. Crome, R. Dawkins, B. Dutrillaux, T. Flannery, A. Gentry, J. Golsen, J. Gowlett, D. Griffin, C. Groves, A. Hamilton, G. Harrington, J. Harris, A. Hill, R. Hinxman, G. Hope, F.C. Howell, W. Howells, E. Huxley, A. Irvine, J. Itani, R. Inskeep, D. Johanson, A. Jolly, C. Jolly, R. Jones, J. Kamminga, J. Kawai, P. Kingdon, K. Kimeu, S. Kondo, M. Leakey, R. Leakey, R. Martin, R.W. May, B. Meehan, H. Paterson, M. Pickford, D. Pilbeam, J. Poole, J. Sabater-Pi, R.I.G. Savage, W. Shawcross, P. Shipman, L. Silcock, A.K. Singhvi, R. Southwood, C. Stringer, A. Suzuki, A. Szalay, I. Tattersall, A. Thorne, S. Tompkins, P. Trezise, E. Vrba, A. Walker, B. Walker, R. White, R. Wrangham, S. Zuckerman.

To Felicity Bryan, Carol O'Brien, Ingrid von Essen, Ronald Clark, and the production team at Simon & Schuster, my thanks for getting this book together.

At various times I have had generous support from the following institutions. The Zoology and Biological Anthropology Departments of Oxford University; CSIRO Canberra & Atherton; the Australian National Museum, Sydney; Makerere University, Uganda; Kenya National Museum; The Mammal Research Institute, Pretoria; Kyoto University and its Primate Research Institute at Inuyama; the Christensen Research Institute in Papua New Guinea; Wellcome Trust, London; the Zoological Society of London; The British Museum and Musée de l'Homme, Paris.

I am grateful to all these institutions and their directors and to many colleagues for shared fellowship and discussion. I also owe some independence of mind and movement to all those who have bought my paintings, drawings, prints and sculptures. Above all has been the support of my immediate family over the years.

At Makerere University my colleague Tag el sir Ahmed pressed me to write an illustrated account of human diversity. He may be surprised by the result but I have attempted to illuminate several chapters with what might be described as 'passport photos', minuscule images that are sufficient to document and identify the various people portrayed. All are deprived of ornament, tattoo and hair so that ancient and Modern, men and women, young and old, share in a similar reduction to essentials. The plates emphasise the variability of individuals as well as the overall resemblances between all Modern humans. These bald passport snaps allow my reconstructions of extinct hominids to be more readily compared with Modern people. These snaps are part of my effort to dig out a lost family album in which I offer decipherable portraits of an African great-grandmother, her Asian daughter and European grandson. My genealogy has artificially contracted generations but it does offer an accessible and comprehensible concept of the human family.

My portrait of this family includes a catalogue of human imperfection and mischief. To recognise such faults is not to accept that opportunism and genocidal territorialism are built into human nature. Hope has better claims to be an intrinsic part of our make-up. A hopeful disposition and an inclination to seek family-hood in our species used to find expression in the old fashioned word 'chierté' (a precursor to 'charity'). Faced with social and environmental crises of ever-increasing severity, an ever more bruised humanity will have to cultivate 'chierté'. Chierté, deliberate, orderly and principled, will be an essential antidote to paranoia, greed, hatred and disorder.

Introduction: A Beginning

What you believed about your beginnings used to begin with where you began.

Way out in the Pacific people have inhabited countless tiny islands for tens of thousands of years. With mainlands long forgotten they imagined their islands as being like fish dredged up from an ocean that was without limit. In the Admiralties, more than 4,000 kilometres from Asia the villagers of Manus visualised the arrival of people as hatchlings crawling from the eggs of a mythic turtle. Out in that all-encompassing sea the spring-tide arrival of giant turtles, rowing in under a full moon was mystery incarnate. For a long succession of generations looking out into the blue from their lonely shore what better symbolic explanation for the islanders' remotest origins?

Ideas about beginnings were no less symbolic in forested northwestern Europe. The Teutonic fatherland, Vigrid, sustained the giant tree Yggdrasil rooted in the centre of the earth, its crown holding up the universe. Around the pivot of its trunk a threatening circle of hostile forces symbolised all that was unknown, unpredictable and unsafe. In the north Fenrir the giant wolf flexed his bloody jaws. From the east other wolves, in packs, advanced in autumn to consume the sun and bring winter. To the west, Midgard the sea serpent stirred up great storms and floods with the thrashings of its tail, while the demon Surt was vanguard to a host of southern fire giants that incinerated Vigrid. For the terrorised Teutons the first humans crept out apprehensively from the scorched heartwood of Yggdrasil.

Everywhere our beginnings have been the stuff of just such legends. Genesis stories have always reflected the experiences and world views of their tellers and in this respect contemporary retellings still suffer from very real and continuing limitations of knowledge. Nonetheless facts have begun to overtake fantasy and the mysteries of origins and of our 'place' in this world are being informed by a rapidly enlarging body of knowledge.

This particular rewrite is a biologist's vision of Modern humans having made a zoological radiation from a region of origin, like many another animal. It acknowledges tradition by embracing Adam and Eve and the

1

Garden of Eden as widely shared symbols to serve as points of departure. In this Genesis an African species, after success in its continent of origin, spreads out (much as hyaenas and lions did) to colonise Eurasia. I am acutely conscious of African beginnings because I was born and grew up in East Africa where, while I was still very young, my imagination was caught by the great age of the world I found myself in. Olduvai, Olorgesailie, Isimila and Nsongezi were picnic and camping sites with a difference! I would visualise the cavalcade of lives that had preceded me there. I would think of those other eyes that had opened on the first flush of sunlight on the hilltops for more than 2 million years, noses that whiffed the smoke of bush fires or scent of acacia blossom and ears that listened to the bush shrikes' duet. The places have changed less than the people who camp there.

Curiously the decision to embark on this book was influenced by a mistake of mine that at the time seemed very trivial. In 1967 when I was writing the first of a series of volumes on African mammals I followed a predominantly American school of thought and used a single scientific name, *Pan*, for both chimpanzees and gorillas. When that volume was published the eminent British anatomist John Napier remonstrated that the 'lumping' of such distinctive great apes called for a protest march on the U.S. Embassy! As it turned out he was right to be incensed and it was two American molecular biologists who finally vindicated his indignation.

I had spent long periods trying to follow, watch and sketch both these apes in their natural habitats and had even taken part in dissecting them (and humans) at the Makerere Medical School. At the time it had never crossed my mind that I might have been more closely related to *Pan troglodytes*, the common chimp, than either of us was to the gorilla. I was therefore as sceptical as most people when Charles Sibley and John Ahlquist announced in 1984 that humans are the closest living relatives of chimpanzees. Nonetheless, the principles on which these and other molecular geneticists based their work have proved sound and their evidence is compelling. Sibley's evidence that chimps are not some forgettable sideline but a living key to our origins (and his molecular clock timing of seven million years for the date of our divergence) has brought an entirely new context to the problem of our biological origins that was first addressed by Charles Darwin. His pioneering search for the

2

forces that extracted us out of a more fundamentally biological matrix than turtle eggs, tree bark or missing ribs remains central to any evolutionary genesis. The new ancillary question is this: how has our biochemical status as just one of an assortment of African apes become so thoroughly disguised?

Few may care to identify technology with the fruit of the Tree of Knowledge but eating this apple had many biological consequences for the path of human evolution: the human form, human diversity, language and our relationship with nature have all been shaped by technology. Humans have become intrinsically different from apes by becoming, in a very limited but real sense, artefacts of their own artefacts.

To take the measure of this proposition put it another way. Traditionally we have assumed that the distinctions that mark us off from other animals are a large ratio of our make-up. Not so: by far the greater part of a human is animal. By far the greater part of what is left is adapted for life in the savannas of Africa. (These proportions are not vague entities, there are over 100,000 but an ultimately *finite* number of genes.) On the balance of this fraction of a genetic fraction the prehistoric life of modern humans has left its mark, and it is here that technology has had its influence. We would scarcely recognise ourselves without it. Those familiar contours of the face in the mirror or on the street evolved only *after* tools and technology had become a decisive force in our destiny.

For instance, it was cooking and processing food that removed the need for massive jaws and teeth and changed the proportions of our faces. It was control of fire that let us invade the cold north and it was building boats that took us to new islands and continents. Such adventures arose from a material culture that was self-made rather than the result of biological adaptation but they precipitated changes in physique that can still be seen in the appearance and complexions of living people.

Humans were born into a man-made world from the very start and the first Moderns took on their environment from a chain of tool-using predecessors. It is a truism that every generation inherits the ruins, the successes and the mistakes of its ancestors. As prehistoric humans dispersed out of Africa their obvious adaptations, such as black or white skins, were made not to climate alone but also to the technologies and cultures that pushed humans beyond their biological norms. Prehistoric habits, born of ingenuity and opportunism, are still an underrated

3

A prehistoric sketchbook. Portraits scratched onto stone in the cave of La Marche, France about 12,000 BP (from tracings made by L. Pales, J. Airvaux and L. Pradel, 1984).

driving force in human affairs. It is with the conviction that there is a continuum and a relevance for the future in our past that the last two chapters are included in a book that is otherwise about prehistory. Biologically we are no different from the people of the late Pleistocene.

This perspective of human evolution views Moderns as the survivors of at least three or four earlier lineages. For every one of these the making of tools, however rudimentary, was intrinsic to the way in which a living was to be made. Each lineage, as it developed, encountered the earlier, less developed economies that had preceded it and every one eventually displaced its predecessors. That much can be inferred from sequences in the human fossil record (which scarcely differ in this regard from other animal lineages).

Viewing humans as a special form of primate carries with it the awareness that every major group of animals has a definable home where millions of years of residence have accustomed them to particular sets of foods, climates, environments, diseases and predators. For the hominids that biological home is the eastern and southern third of Africa.

Today, genetics and biology are as important a part of reconstructing evolution as palaeontology, and new discoveries in all these fields of knowledge follow fast on each other's heels. Earlier accounts of human evolution have been desperately dependent on making connections between a precious few and very incomplete fossil bones. Not only are there many more bones now, but the awareness of just how varied Modern populations are has devalued single fossils as 'types' or models for past populations. There are now long series of fossils, all from single sites and of the same period, to demonstrate that early Moderns were as varied as, and possibly even more varied or polymorphic than, their equivalents today. There is also a suggestion that prehistoric communities could have been predominantly heavy and robust when populations were small and scattered, but lighter and more gracile when populations were dense and numerous. The admission that classically 'primitive' or 'more advanced' types may reflect densities or some such simple gradient could greatly change the way evolution is viewed. Instead of being a temporal progression from heavy to light, these extremes may have alternated, and the preponderance of one or the other may have ebbed and flowed over time. Only very recently have wholly gracile people come to dominate the scene. Could it be mere coincidence that densities

5

and numbers have risen sharply over this period? This possibility is a very new concept and one of its implications is that wholly modern genes might have sheltered within some very un-modern-looking skeletons.

As our ancestors colonised new areas and new habitats, they not only developed new techniques and discovered new foods, they also adapted. Adaptation meant the selective survival of those individuals with the most favourable genes. Sometimes adaptation involved physical changes that came to typify a whole regional population, but other characteristics could have been no more than a product of the passage of time and isolation. There is still much sorting to be done to distinguish functional from nonfunctional genetic differences. These changes, which might involve alterations in body proportions, the shape of features or the tints of eyes, hair or skin, were made easier by the wide range of variation *already* typical of humans. Modern humans have achieved a great deal of genetic variety in spite of their evolutionary radiation being very recent.

A vivid picture of prehistoric diversity has been found in the rock art of a French cave that has been dated at over 14,000 BP. The Magdalenian artists of La Marché were unsophisticated but some of their scratchy drawings would pass as portraits of people we know. The range of fat, thin, tall and short, long and snub noses, high and low foreheads is astonishing for the inhabitants of a single cave. Their likeness to us is confirmed by the fact that Magdalenian skulls are indistinguishable from modern ones. Across the Mediterranean, at Mechta Afalou, in Algeria, a comparable range of types of similar age turns up in a single stratum holding nearly fifty skulls. Select just four of these for comparison (see p. 8). Quite startling differences are apparent – round, square and pointed faces, narrow, broad and hooked noses; flat narrow foreheads, high domed ones, big brains, smaller brains.

One of the most persuasive indications of how much individuals differ from one another comes from immunology. When an organ is to be transplanted there are huge odds stacked against it, because there are so many specific peculiarities in the biochemistry of an individual. These far outweigh the generalised differences that characterise the so-called racial groups.

Among several superannuated models of human evolution are those claiming that several of today's continental 'races' descended independently from different pre-Modern ancestors. Such theories saw prehistoric societies as geographically static, made too much of racial differences and sought to stress them. Since earlier forms of humans were significantly less developed than Moderns, there were also false implications of different capacities and different potential between 'the races' in such models. These theories have been largely abandoned by later anthropologists and are not given credence in the pages that follow. Like the great majority of my colleagues I judge the evidence as pointing to all Moderns being of recent origin and closely related. Nonetheless the very obvious regional differences, commonly categorised as 'racial', do call for more explanation than mere 'natural variation'.

If both black and white skins originated from 'self-made' climate ordeals, 'whites' are depigmented ex-Africans. As for 'blacks', origins are much more complex. The image of a monolithic black population stagnating behind the Sahara must be rejected as a false stereotype manufactured in Europe more than a century ago. The parent stock of all Modern humans continued to change and diversify in Africa just as their diaspora did in Eurasia, Australasia and the Americas but there was an added twist. Africans have had extremely ancient coastal connections with oceanic Asia.

The landmarks or beacons that mark out the Afro-Asian seashore connection are scattered through space and time. All depend upon some form of material evidence; sometimes it is fossil skulls, sometimes a scatter of artefacts. Some of the evidence is based on gene patterns or physical differences in living people. Connecting these isolated components is the diffusion of a single lineage of Modern humans out over real lands during real time. The clues to reconstructing that progression and its dating are still very tentative and scattered: yet they are sufficient to provide a framework for this book. Five chapters retrace those travels, each time from another perspective – biogeography, palaeontology, archaeology, ecology and genetics provide these different ways of viewing a central subject, 'Self-Made Man'.

In common parlance there is an implication of seeing and seizing opportunities in the success of a self-made man. This opportunism is pursued in the final chapter, where it is argued that humans used their technology, from the start, in ways that took little heed of its effects on other animals, on the plants or indeed on any other part of their

Four skulls from Mechta Afalou and sketches reconstructing their general appearance.

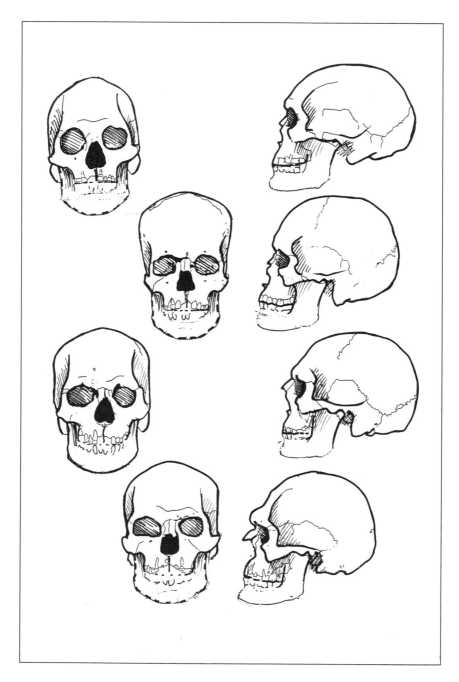

environment, least of all on their own descendants and future generations. I have also taken up an observation on technical innovation that
I first pointed out some twenty-five years ago:

> Progress derives from man's employment of tools, of true speech and from his
> tendency to migrate and it depends upon a more or less continuous human
> tradition, yet with every step of progress man's history, his origins, his earlier
> environments are annihilated or obscured. Almost everywhere reminders of
> the past embarrass the myths of the present. The capacity to make things, to
> develop them and then make them obsolete is a central characteristic of
> human traditions and human behaviour. . . . As a dominant animal, man
> replaces the other wild animals wherever he is able, turning over to his own
> use the natural resources which the animals formerly used.

The dangers of annihilating the past are very real and the search for
human origins is fuelled by more than simple curiosity. Plotting our
history and our ancestors' movements and their doings are the first steps
in reconstructing a sort of lost corporate memory. We need that memory,
because without it we are like amnesiacs, who repeat their mistakes
every time a situation recurs or relive their grief with every reminder of
a relative's death. Learning to cope with the complexities of an already
over-complex life is not helped by forgetting or by being ignorant of their
less complex origins.

Some clues are to be found in the societies of various surviving
relictual peoples, all of whom are under massive (and often grossly
unjust) attack from more powerful neighbours. All are, or were until
recently, hunters and gatherers. The San of the Kalahari, the Andaman
islanders, Filipino and Malaysian Negritos are examples. Before these
distinctive peoples and cultures disappear within the churn of contemporary life there are innumerable traces to be retrieved of travellers'
sagas, of a heritage that is common to us all: our long, long prehistory as
foragers.

In describing the steady expansion of foods that humans might have
eaten as they enlarged their range and their technical repertoire I have
been selective, focusing in some detail on key staples such as the meat of
large animals, clams and yams. While human foods can be described and
understood as nutrients in the same way as those of any other animal,
such a restricted perspective will fail to do justice to the religious and

social connotations that dominated prehistoric perceptions of food. Just as Christian faithful of today implore God to 'give us this day our daily bread' so did earlier people feel the need to thank, cajole, sacrifice to and beg the hidden forces that could withhold or provide their food.

The link between foods and techniques is an intimate one as much in biology as in archaeology. Of course humans are not the only animals that adapt to conditions of their own making. Many social insects husband their food supply and create their own self-contained eco-systems. For example wood-harvester termites lay out fungus gardens beneath elaborately constructed air-conditioning ducts, build a queen's royal chamber, crèche bunkers and lots of connecting passages and foraging tunnels. Within these wholly artificial 'cities' termites have evolved castes of odd-looking drones, queens and workers that could never survive outside the termitary. They too could be said to be 'self made'. Mice, birds, insects and spiders weave the most ingenious nests, webs, fabrics and traps just as others can bore circular holes in wood, burrow in soil, use stone or wood as hammers or anvils (some vultures, parrots, otters and mongooses do), wield branches and probe with twigs like chimpanzees. But, unlike almost all these examples, humans do not restrict themselves to just one or two genetically fixed techniques.

The evolution of technology in prehistory may have proceeded by fits and starts but it has a recognisable theme and direction. The adaptation of technology mimics adaptation in evolution, in that every improvement or refinement of a tool enables it to proceed faster or take in a wider arc than its predecessors. It is commonly said that early humans broadened their niche; instead it might be truer to represent the invention of each new technique as the acquisition of a new niche and its addition to the others already possessed. The effect of this multiplication was to exclude and eventually replace a succession of other species.

One of the incentives for writing this book was to explore when, how and where the world became 'home' (the *Oikos* of ecology) to humans. I also wanted to explore the antithesis to this idea: to what extent humans have been vandals in the home from the start.

Seeing what a large troop of baboons, a herd of elephants, or a locust swarm can do to the countryside leaves no doubt that any animal can be immensely, if momentarily, destructive. The fundamental difference between them and the human animal is the latter's possession of technology. In a climate of much sentimentality about 'natural people', I have sought to explore the way humans *are* technology. That is what distinguishes us and always has. To be 'at home' in this world has long been impossible without some level of technology, and the duality of home-maker and alien is present in every one of us. Nonetheless, if prehistoric trends are projected on into the future, the unimpeded progress of our technology will consume and convert every possible resource, animate and inanimate, to service our ever-enlarging appetite.

The prehistoric perspectives of this book have put natural limits on its scope. Nonetheless, its global viewpoint and identification of tools as the perennial interface between humanity and nature has led to at least one stark conclusion. Today's exploiters of the world's remaining natural resources are generally less knowledgeable and less responsible than their prehistoric counterparts. They have incomparably more powerful tools. Their expertise is seldom relevant to the central task of ensuring that resources are self-renewing. In every respect technology *must* become subordinate to organic life processes and to ecological principles. In other words technology is *not* for technicians to control.

If there is an earnest will to understand the many processes – physical, biological and cultural – that have brought humanity to its present condition, then there may still be the vestiges of a recognisable *Oikos* to live in. It needs to be a home where we can see clearly that from which we came and of which we are still a part.

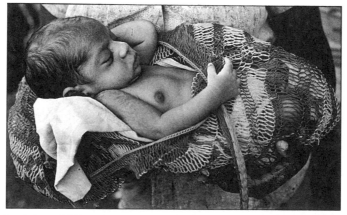

Newly born child in string carrier bag, or Bilum, Madang, Papua New Guinea.

1: Before the Wise Men

Three-and-a-half million years ago at Laetoli in northern Tanzania there was a local disaster. Carbonatite ash and tiny globules of lava had rained down as the Sadiman-Lemagrut volcano erupted and the Rift Valley slopes below were powdered with a sort of raw cement. Then it rained. Such events would have been common enough along the Great Rift and fatal, or at least frightening, for those who witnessed them. We know there were witnesses, because before the mushy cement set into a hard pavement a female hominid, a southern ape woman, or *Australopithecus*, and her youngster trudged through it, probably seeking to escape a suddenly poisoned homeland. After their passage a three-toed horse went by and a rather confused hare dithered in the noxious mud.

All along the tortuous path that leads back from us to ever earlier ancestors were people who had to fill their stomachs and with exquisite spasms of sexual chemistry pass on their genes. Time and again they faltered, as drought, poisoned ash and a multitude of hazards conspired to destroy their frail substance. Fossil bones and footsteps and ruined homes are the solid facts of history, but the surest hints, the most enduring signs, lie in those minuscule genes. For a moment we protect them with out lives, then like relay runners with a baton, we pass them on to be carried by our descendants. There is a poetry in genetics which is more difficult to discern in broken bones, and genes are the only unbroken living thread that weaves back and forth through all those boneyards.

African prehistory and palaeontology are new sciences. At a time when the value of fossils is taken for granted it is easy to forget that human fossils remained virtually unnoticed until Darwin created the scientific and philosophical framework in which they could find relevance. Before the interior of Africa had been explored and with no relevant fossils to hand, Darwin wrote:

> It is probable that Africa was formerly inhabited by extinct apes closely allied to the gorilla and chimpanzee and that these two species are now man's

13

nearest allies; it is somewhat more probable that our early progenitors lived on the African continent than elsewhere.

This farsighted prediction was the result of the framework in which Darwin ordered his observations and deductions. That framework is now the basis for biological teaching and for our understanding of human prehistory. There are now many more hard facts to learn but Darwinian leaps are still necessary to bridge the unknown spaces in between.

Histories cannot be entirely taught. To perceive history as in any sense a living past, rather than a procession of learnt facts, requires feats of imagination that are essentially private and voluntary. For imagination to be more than fantasy, our experience of the living world must offer some sense of continuity to help bridge the chasm between that poisoned day at Laetoli and the present. Three-toed hipparions have gone but a barking zebra signals some sort of equine continuity. The sun that rises over a now extinct Lemagrut will never again illuminate the tread of an *Australopithecus* family but the genes that could build two flat feet like theirs are not extinct. Toe by toe and heel by heel there are countless feet being built in countless wombs today that at the right age could retrace those trails across volcanic mud.

With the recognition that genes are the real continuum it becomes less important whether we are direct descendants of that hominid woman or not. It is not even important whether any fossil yet found has left descendants. What is important is that the genes that build living feet and hands, and seeing, listening, eating, thinking heads, today have a continuous lineage back to the time when big toes were thumbs and arms seldom hung free but propped up a heavy cantilevered chest. The genes that still build the splendid edifice of a gorilla or the rumbustious form of a chimpanzee have no closer structure than that which builds a human. At the molecular level I differ from a chimp by only 1 per cent. I am closer to the chimpanzee than it is to the orang-utan.*

*Any special status that I may claim over the chimpanzee is not due to an enlarged genome. I have no new genes, only *modified* ape ones. At the molecular level I am less different from a chimp than a mouse is from a rat. My haemoglobin is identical to the chimp's in all 287 units. Of 1,271 amino acids I share 1,266 and my genetic distance using the DNA hybridisation method only differs by 1.6 per cent. Interestingly the two chimp species differ in their DNA by as much as 0.7 per cent. One measure of our *differences* against all these resemblances lies in calculating the number of nucleotide positions involved. As two specimens of *H. sapiens* you and I differ by up to five million positions, most, but not all of it,

(left) Schematised diagrams of homologous chromo-
somes (7/8) from human, chimp and gorilla
showing greater similarity between the first two.
(right) Diagram of more bands in another
chromosome (11) showing detailed resemblance
between humans and chimpanzees.

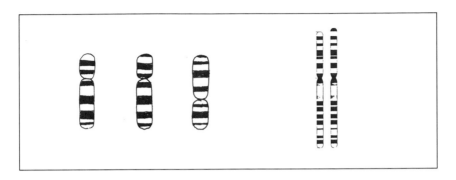

The great African apes are genetic Rosetta Stones whereby we get the measure of our humanity. Whatever happened after that point at which we parted company, perhaps as recently as 5 million years ago and certainly no more than 8 million, is the short human story. The apes, instead, continue the main line of a much longer and older story, one that can be retrieved only by feats of patience and humility and observation in the few and shrinking forests where apes have learnt to trust humans. For example, recent observers have recorded very local-ised traditions among different chimp populations. In the Tai forest, in Côte-d'Ivoire, the chimps find hammer stones and carry these, together with hard nuts, to tree-buttress anvils where they smash open the nut shells. Distances of up to 500 m are travelled. By contrast at Bossou, in Guinea, they carry stone anvils as well as round hammer stones right to the site of the nut fall. In Zaïre chimps use staves, in Tanzania they fish termites with twigs and leaves. The Spanish anthropologist Sabater Pi has pointed out other examples of locally restricted traditions which he regards as primitive regional cultures. Especially important about these behaviours are the elements of anticipation and planning. In the jargon of one school of anthropology, they show 'time planning depth in their technologically-aided, task-solving food-procurement endeavours'!

Not only is it known that the chimpanzee lineage is much older than our own but there is also a genetic measure of their age. They have been around ten times as long as us because they show ten times as much

---

molecular junk. Either of us differs from a chimpanzee by about fifty million positions which represents about 1.5% of the total DNA sequences.

genetic variation, confirming that all Moderns derive from a recent common ancestor.

As hominid footprints prove, the first novelty of our lineage was the way we walked. If the divergence was indeed as recent as 5 million years ago, the genetic changes that pushed us upright were fast. The first savanna apes are likely to have been generalists of a similar type to modern chimps. The latter combine a largely vegetarian diet with occasional forays into active hunting and meat eating. Only after becoming a successful bipedal type of hominid did the Australopithecines diverge into at least three clusters. The lightweight Lucies (*Australopithecus afarensis*) had thick enamel and low blunt cusps to their teeth, characteristics that anticipate *Homo*. The Nut-crackers, *Paranthropus*, evolved big heads with heavy reinforced teeth in powerful jaws. The third type, the original 'southern ape', *Australopithecus africanus*, was intermediate and may have been a conservative relative of the Nut-crackers. It seems to have been less successful as it is only known from South Africa.

While savannas have a rich variety of foods including many fruiting and seeding plants these resources are more widely dispersed than in forest and much more subject to seasonal fluctuations. The divergence of australopithecines into separate lineages represented three responses to this last element of irregularity. For example, the response of the lightly built southern apes may have been to enlarge the total range of territory individual animals could cover. The Nut-crackers instead adapted by taking on the lower-quality foods still available in the dry season. The Lucies' strategy is less clear-cut. Their teeth imply more seeds and nuts but they are also likely to have become more omnivorous.

That omnivory offers many real advantages finds confirmation in the great number of omnivores that still flourish in African habitats. A majority of the smaller African carnivores combine meat, carrion and invertebrates with fruit-eating. Some, such as genet cats, have radiated into many species and races which differ in the relative proportion of these foods in their diet. The palm civet relies on fruit as its staple but takes animals or carrion whenever these are easy to get. By contrast those species with a mainly animal diet, such as mongooses, civets, striped hyaena and jackals, may abandon the hunt when there are seasonal gluts of fruit. The choice of an omnivore's staple is often a trade-off between the animal's anatomy, its feeding techniques and the ease with which a particular food can be got at any one season or place.

The main anatomical difference between the earliest *Homo* and the Lucies was brain enlargement. The first appearance of stone artefacts implies that the main difference in feeding technique was the employ-ment of tools. An alliance between brain and technique would not only have enlarged the overall choice of foods but also implied an overall decline in the importance of fruit. There is much dispute, however, about the pace, degree and timing of that decline in fruit and plant eating. To visualise the capacities of emergent *Homo* we do not have to rely on pure imagination. Modern chimpanzees drive, panic, block, divert and am-bush smaller prey (mostly monkeys). In playing specialised roles as drivers, sentries or killers, they display some of the strategies used by social carnivores such as lions and wolves. They also appear to use vocal and visual signals, which probably help to coordinate hunters in the highly obstructed forest setting and thus can 'direct' the sequence of attack and capture. These bouts of hunting are very energetic and are seldom routine, and their effectiveness could probably improve in a less difficult, more open environment. The efficiency of their technique could also improve through practice were the frequency of their hunting to increase.

Only in the generation of a more coordinated pattern of attack and capture could primates hope to out-compete other predators in the savanna. Some of the greatest gains would have lain in improving the system of signs and signals that enable any hunting group to anticipate, control and guide a prey animal's behaviour so that it can be more easily caught and killed. This is not to say that the rather small omnivorous australopithecines were formidable big-game hunters, but there are many small animals susceptible to such methods. The spoils of commu-nal hunts are shared by chimpanzees, so the more frequent and more effective efforts of australopithecines would have rewarded more par-ticipants. Since rewards are the incentives to improve any performance, there were the makings here of an entirely new type of predator.

Another simple evolutionary pattern concerns endemism. African apes are endemic to equatorial Africa in the same sense that kangaroos are Australian and toucans Amazonian. Endemism can be more specific than that; well-defined subregions and ecosystems spawn their own clusters of species, each with a more refined niche of its own. Thus, wild *(left) Schematic map of centre of endemism for African forest monkeys (Zaïre basin). Shows the main zones of expansion for the more terrestrial savanna and patas monkeys and for gentle monkeys (common in eastern gallery forests).*

sheep and goats have proliferated in the Himalaya Mountains and their outliers, while the dwarfed dik-dik antelopes have diverged into five species in the arid Horn of Africa.

To take mammals that have been the subject of my own field studies (and are also African primates), the guenons are archetypal tropical African endemics. The centre of evolution for these long-tailed monkeys is the Zaïre basin. From this central complex of forested rivers, guenons have spilt out in all directions to colonise drier or more marginal habitats. The more terrestrial types have spread furthest, while the most arboreal remain confined to the forested equatorial heart of Africa. Under the influence of oscillating climates the guenon prototypes in Zaïre have thrown up a succession of refinements. Some have become smaller and more insectivorous, others larger, eating more foliage; one is the seden-tary gleaner of a tiny home range, another a more opportunistic traveller over a very large one. Not all are equally successful and the more conservative types are among the rarest and most restricted in range.

In Southeast Asia, the guenons' equivalents, macaques, have seen a similar radiation with many forms evolving at the heart of their range, while a few, hardy and successful generalists have colonised woodlands as far away as Japan and the Mediterranean.

These patterns are relevant to understanding human evolution, be-cause they show that the region or ecosystem in which a successful adaptive type first emerges remains significant long after descendants have spread more widely. Gorillas, chimps and pygmy chimps, or bonobos, survive not just because they are adapted to 'forest' but because the much smaller enclaves of forest in which they now remain are their evolutionary heartland. Here, in spite of awful obstacles, they persist within their original centres of endemism in western and central equatorial Africa.

It is no accident that the hominid centre of endemism lies immediately east of that of the apes. This is the geographic and ecological dimension of the hominids' adaptive shift into drier zones.

To understand conditions in the past, models have been constructed to show (very approximately) where forests would grow if today's rainfall increased or diminished. The distributions of many forest ani-mals and plants show some agreement with this model. More rain would cause expansion in all directions but the mountains, plateaus, rifts and *(right) Effect of forests 'pulsing' under the influence of climate change. Simple expansion/contraction in west. Isolation + degradation in east. From a primary zone in eastern Africa savanna apes are likely to have diffused into distinct extension zones to the north-east and south-east.*

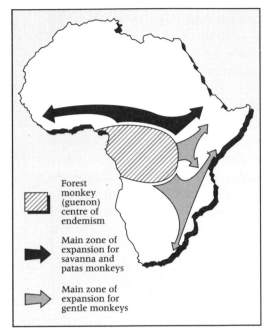

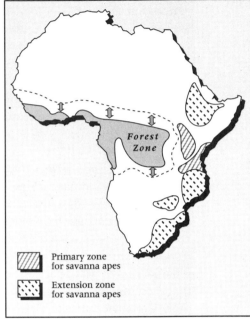

Forest monkey (guenon) centre of endemism

Main zone of expansion for savanna and patas monkeys

Main zone of expansion for gentle monkeys

Primary zone for savanna apes

Extension zone for savanna apes

rain shadows of eastern and southern Africa fragment the pattern. A distinct zone of eastern littoral forests and forest mosaics is sustained by Indian Ocean rain clouds (which behave independently from those to the west). This fragmentation pattern invites a surmise as to where the first eastern ape populations might have been. It also suggests the pathways of later secondary expansions by successful hominids.

There are no physical interruptions north or south of the Zaïre basin's central forests, so ape populations would simply ebb and flow with the climate without the pressure to adapt to non-forest. Further east, there were barriers to break this easy coming and going. Then as now, the Western Rift and its associated mountains and lakes formed a natural, albeit permeable, obstacle course.

Because of their equatorial origins the first eastern apes would have inhabited only the lands now called Tanzania and Kenya (minor southerly extensions might have intruded into today's Malawi and Mozambique). Eastern and southeastern forests would have clumped in four main blocks separated from one another by dry corridors. These blocks

19

might have been tentatively and imperfectly linked during the peak wet periods but the distribution of many archaic organisms reveals that each of these blocks has retained its identity as a biogeographic unit through the driest and the wettest periods. With each dry pulse, the forests within these large blocks would have degraded, fragmented or even disappeared as recognisable forests, but many of the hardier species originally associated with the moister conditions have continued to survive tied to their regional centre of endemism.

Since the four forest-growing zones are strung out on a very long north–south axis (with many narrows), the early hominids that eventually colonised the eastern third of Africa would have tended to get subdivided in a similar fashion. Alternating climates might have made the preferred environments of early hominids oscillate but their central cores are likely to have followed the eastern Rift Valley. Since populations were strung over such a long and often narrow chain, the evolution of distinct regional types would have been a natural outcome. These regional centres can be identified with some confidence, and it is likely that these subdivisions retained an enduring significance for much later, more advanced human populations.

An accessory to eastern Africa being the 'home' of hominids is the likelihood that the more important innovations (as for the guenons in Zaïre) evolved there. After several million years of occupation, hominids have become more precisely tuned to local conditions here than to anywhere else. Unlike more distant colonists (which have the challenges of different climates, foods, diseases, competitors and predators to adapt to), those in the centre of endemism can 'refine' the central traits that are most typical of their lineage. This may help to explain why eastern Africa seems to have spawned about five distinct clusters of hominids, each representing a major advance in human evolution.

The first shift began between 5 and 8 million years ago, when recurrent periods of aridity would have forced many marginal populations of forest animals to adapt to a slow but relentless degradation and isolation of their environment. For eastern populations of bonobo-like apes, the drying out would have involved some predictable changes in behaviour. From being one of the larger and more mobile members of a community in which primates dominated and the trees gave instant safety, they became (relative to their new neighbours) small-sized, vulnerable and rather marginal members of a crowded community that included many carnivores. Of course, they would have brought with them an already superior intelligence, a tightly coordinated group structure and the habit of moving at a fast pace between spaced-out concentrations of fruit. With fewer and more widely scattered sources of food in a more dangerous setting, the most immediate pressures would have been to increase the speed and efficiency of movement and improve defensive behaviour. With more time spent on the ground, and moving at a faster pace, entirely new demands were made upon the foot. It now had to be better at fast walking than slow climbing, so only two avenues of change were open. A now obstructive and heavy thumb could get out of the way by declining in size or it could align itself with the other toes to add its strength to a single strut (in lieu of the originally powerful forked clasp).

The choice of aligning all the toes might have been influenced by the apes' tendency to stand and monitor the surroundings. This would have helped develop stout hind feet capable of taking the entire body's weight, but the maintenance of a permanently upright stance followed rather than preceded the foot's reorganisation for fast terrestrial movement. The changes might have been relatively rapid and the primary incentive for becoming bipedal was perhaps a simple matter of energetics. Given some reorganisation of muscles and bones it can be as economical of energy to stride on two legs as to run on four. Tests on young chimpanzees have shown that they are energetically just as efficient walking on two legs as on four. To envisage or describe the ancestral ape as a quadruped is misleading. Some apes do habitually carry part of their body weight on the forearms but, compared with a true quadruped such as a horse, the 'forelegs' are awkward contrivances. For an animal lineage that once brachiated up in the tree-tops it was probably easier, anatomically, to widen the separation of function between fore and hind limbs than to narrow it. To force a branch-swinging arm to become a foreleg and forgo some of the manipulative skills of a hand was a weaker option. The usefulness of hands and arms for exploring and collecting potential foods is greater for an omnivorous opportunist than for a more conventional herbivore. The preference of all apes to get up on their hind legs while making a display could also have discouraged quadrupedalism. If erectness was an evolutionary response to the demand for a more sustained and faster pace it is less likely that the posture was adopted in

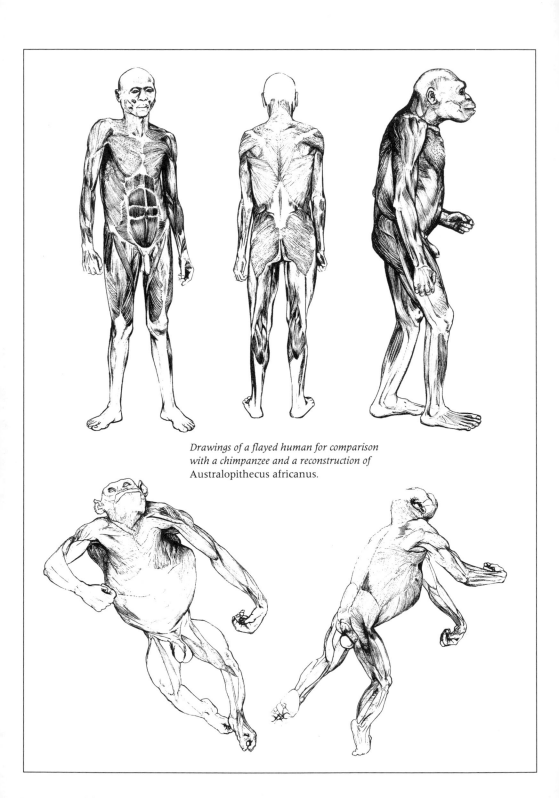

Drawings of a flayed human for comparison with a chimpanzee and a reconstruction of Australopithecus africanus.

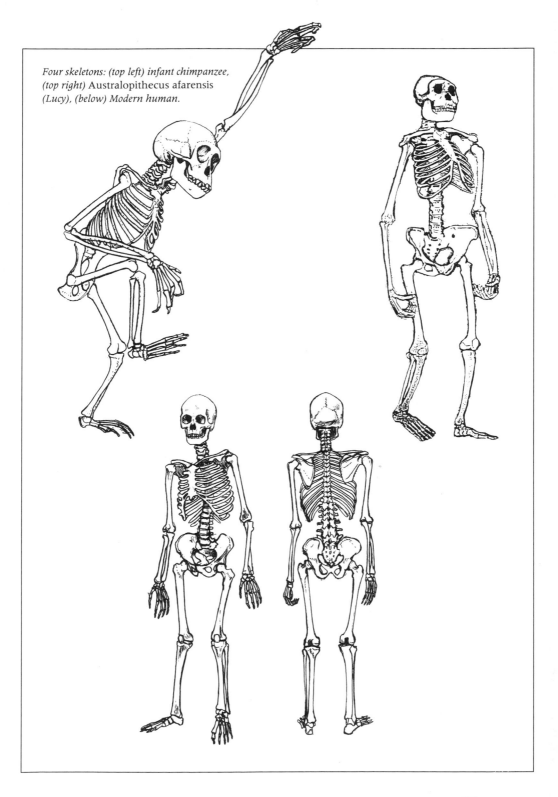

Four skeletons: (top left) infant chimpanzee, (top right) Australopithecus afarensis *(Lucy), (below) Modern human.*

23

Bonobo proportions are distinctive. They are longer in the leg and more slender than common chimpanzees (sketches from photographs).

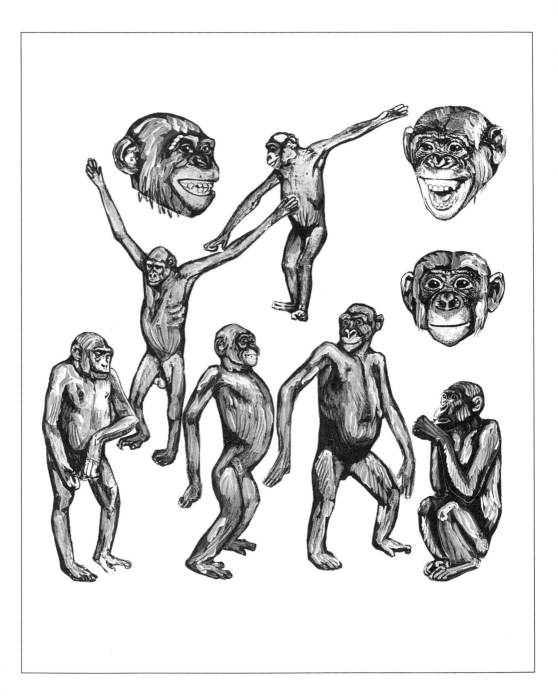

true forest than on less encumbered and more danger-ridden ground.

There was a significant lag between becoming habitually bipedal and reorganising anatomy to support the new gait. Australopithecines carried themselves upright like humans, but they were still bonobo-like in some of their muscle attachments, and the pelvis in particular was intermediate between ape and human.

Some early students of origins faced with rather few fossils of uncertain dating modelled human evolution as the simple linear progression of a single lineage. In the species-saturated savannas of Africa, it seemed inconceivable that there might be room for more than one hominid species. This was a mistaken assumption, because ecological niches proliferate rather than contract in a diverse community. In rich and varied ecosystems plants speciate because of all the physical, climatic and biological variables – as opportunities and pressures multiply, so do species as they respond to change. Likewise, herbivores specialise more and more in their food and feeding strategies. Predators, scavengers and parasites develop ever more refined and ingenious ways of catching and processing their prey or penetrating their host. In such a crowded and dynamic community, successful ecological niches develop around unique *techniques* of making a living.

Of the three African apes, the bonobo, or pygmy chimp, is most prone to an upright position. It differs from its close relative, the common chimpanzee, in a number of significant ways. It has a distinct karyotype (or difference in chromosomes). It has a rounded top and back to the skull, short canines and a longer femur. It forms mixed rather than single sex groups, is shy and highly strung and makes a high-pitched rather than a hooting call. It has been found to have a markedly different physiology. Females have a less well-advertised oestrus cycle and their vulvae face more to the front than in other apes. Coincident with the latter differences is an interest in sex that has endeared them to more than one primatologist:

> [Bonobos] seem to be interested in sex in all its manifestations and variety: dorsoventral and ventro-ventral, heterosexual, homosexual, month round, as part of food-sharing activities or as a part of normal social interaction, or for no particular reason at all ... they make spontaneous hand gestures to one another indicating where and how they want it. (Groves, 1986.)

25

Bonobos also occupy a relict distribution in the southern Zaïre basin. This is an area where most of the forests would have shrunk during the driest periods into snaking galleries that lined the diminished watercourses. Today many organisms in this area are true forest dwellers, yet they have very close relatives living in fragmented savannas in East Africa (elephant shrews, bush rats and hard-nosed toads, to mention but a few). All these species, including the bonobo, are currently in a 'full-forest phase', but it is certain that they have all survived several periods of much greater aridity in the past. This history of instability in their southern Zaïre basin environment may have influenced the subtle ways in which the bonobo differs from the common forest chimpanzee. A special interest lies in the bonobo's possible built-in capacity to weather repeated changes in its habitat wrought by climatic change. Nonetheless, it is because it has never been detached from the main forest block that the bonobo remains emphatically a type of chimpanzee and not a hominid.

Populations of similar animals in East Africa, physically and genetically cut off from their parent forests and from the central forest gene pool, would have made less subtle changes. The degradation, fragmentation and transformation of forests was both more complete and more permanent than in the west, so the bonobos' adaptability would have been transformed into actual adaptations.

Once these adaptations were effective the eastern apes, now the earliest hominids, should have become widespread and typical members of the East African savanna communities.

In the broadest sense all African apes are relicts from a time when apes were abundant and diverse. All the survivors live in forest, which has been one of the last habitats to be occupied or attacked by humans.

Eastern Africa has always been a broad corridor of savanna and woodlands well laced by lakes and rivers, valleys and hills. It is here that the 'southern apes' probably emerged as a more bipedal eastern population of a very common lightly built ape. Considering how superficially similar chimps, gorillas and orang-utans are today, an equally superficial glance at that animal might well have described it as an unusual form of bonobo. The latter is certainly the best illustration we have of the sort of animal from which humans could develop. Given that there is still a vital living population of them, it could be said, in the present context, that

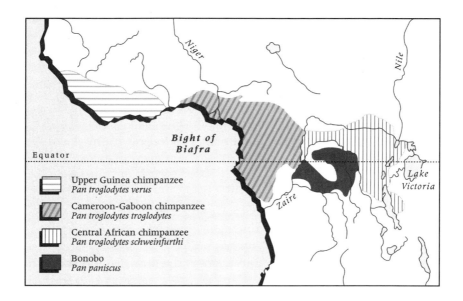

they are the single most important survival from the African past that exists. Their survival is more important than any fossil, but the reluctance of most humans to accept their animal origins consigns the bonobo's fate to an advancing tide of peasant farmers and to the whims and greed of loggers. Even so, its survival as a species distinct from the more widespread forest-dwelling common chimpanzee is also a token of its extraordinary vigour and distinctness. Were it more like the common chimp it would have hybridised itself out of existence. Were it too peculiarly specialised, it would have succumbed to the vicissitudes of numerous climatic extremes and habitat changes. It may be a relict but it is the very persistent and tough relict of what was once a more widespread and abundant animal.

Resilience and abundance – these were likely attributes of the first apes that moved into the savannas. In a highly diverse and constantly changing continent, these were essential properties for rapid speciation.

Four fossil species of australopithecines have been agreed on so far, but it is likely that each of the four types had quite distinct subspecies and that any one of these had the potential for further evolution. Intense competition forced some populations to specialise in eating foods that

Tracings from photograms. (a) Distribution of weight by Laetoli hominid foot when walking through wet ash. (b) Modern human foot. Outline of foot bones in (c) chimpanzee, (d) Australopithecus *fitted to a Laetoli footprint (dotted outline), (e) Modern human. Note different proportions of big toe and third phalange.*

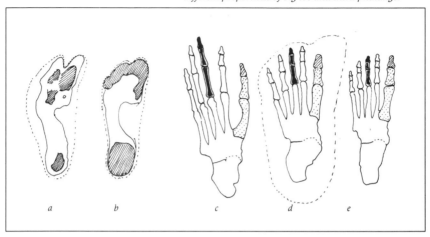

were abundant but difficult. Thus, hard fibrous plants and the very toughest of animal food may have become the speciality of several big-jawed heavyweights, the most famous of which was the 'Nut-cracker man' or *Paranthropus boisei*. At least three species have been described and these specialists survived much longer than any of the more generalised earlier forms. Of the latter, the earliest known so far is *Australopithecus afarensis* which was possibly similar to the mud-walkers at Laetoli. This species combined the same brain size as a bonobo and a similar skull form with fully erect posture and a modified set of teeth.

The first complete skeleton of this hominid was found in Ethiopia and was nicknamed 'Lucy', because the young palaeontologists who found her were Beatles fans. 'Lucy in the sky with diamonds' on the camp tape recorder expressed their jubilation at finding a 3.5-million-year skeleton of a hominid (at that time the earliest) on, or close to, the main line of human evolution.

The population Lucy belonged to was also quite pygmoid and delicately built. Other fossils allocated to the same species were much larger, and widely divergent body sizes from region to region and time to time have remained typical of hominids (including Modern humans) as they have for many other mammals.

The prints that were left in the Laetoli mud so long ago may be almost indistinguishable from those of modern feet but, now that Lucy's feet (and those of others of her kind) have been examined, it has been found

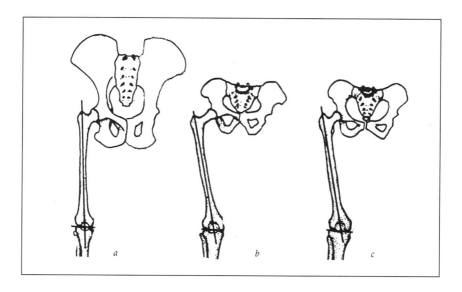

that there are some subtle but significant differences. The contours are indeed human but the ankle joints were still chimplike in their flexibility (something more necessary for tree climbers than walkers and a clear indication of origins). The toes also offered hints of their arboreal derivation in being exceptionally curved and strong.

Lucies also had very human hands but the thumb was still small and the middle finger was rather sharply tapered. Nonetheless, both hands and feet had changed in a predictable way. Feet that no longer had to grip branches at odd angles became simplified into a compact arched pad which had lost much of its versatility. The hand, by contrast, increased its flexibility, because it no longer had to combine manipulation with support. Knuckles freed of weight-bearing could slim down and an over-trim thumb could emerge from its subordination. In the course of further evolution it became the strong digit that grasps this book, this pen, a knife or a flint.

These and other details of the joints and hips reveal that a four-legged, tree-climbing architecture had been modified into the sort of forms that we recognise as typically human – except for the head. When the first incomplete juvenile skull was described in South Africa in 1925, opinion was sharply divided on whether it was just another ape or a hominid. Critics of the hominid claims were not being unreasonable in their

*The heads of southern apes
(Australopithecus and Paranthropus)
looked more ape than human. Not their
bodies which were already erect three million
years ago: (a) Robust nutcracker,
Paranthropus robustus (profile and
frontal), (b) Southern ape, Australopithecus
africanus (frontal and profile), (c) Bonobo,
Pan paniscus and three views of Lucy, (d)
Australopthecus afarensis. Reconstructions
of savanna ape: lucy, southern ape, nutcrackers
(boisei and robustus).*

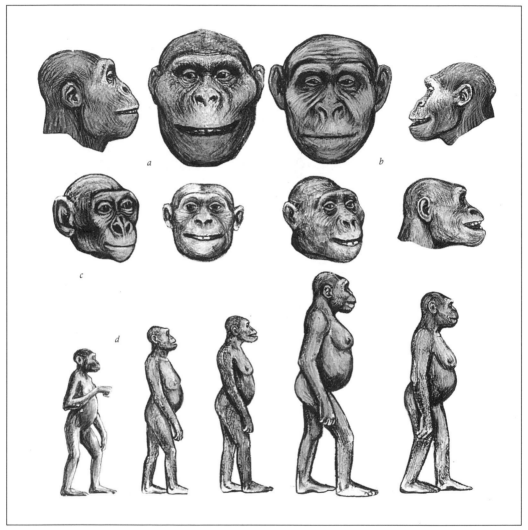

insistence that *Australopithecus* heads were more ape than human. After all, changes in dentine and tooth structure had some precedent in the orang-utan (which has a more humanoid, thicker enamel than the African apes). Furthermore, their brains were proportionately of the same size as chimpanzees'.

Australopithecus could no longer be dismissed as a mere southern ape

Upper jaw tooth rows: (left) Chimpanzee, (centre) Lucy, (right) Modern human.

BEFORE THE WISE MEN

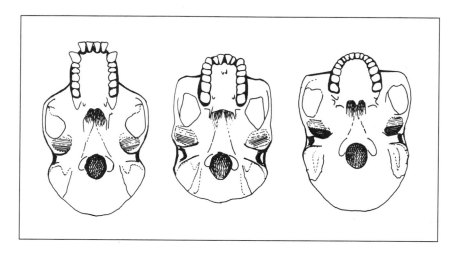

when, many decades later, Lucy and other skeletons began to be unearthed in ever greater numbers.

I have stressed that it is unimportant whether Lucy or any other *Australopithecus* are on the same main line of human descent, not to minimise the significance of such finds, but to shift the focus off individual animals and on to hominid ecological niches in Africa. The five or more australopithecine species that have been identified so far represent a mere sample of what were probably the more common species. Being from localities that favoured fossilisation is no guarantee that they are fully representative. Indeed, it may distort the picture by inadequately representing the moister tropics.

There tend to be two polarities in animal adaptation. There are the extreme specialists that do one thing supremely well. Thus, anteaters, fishing bats and lily trotters, for example, tend to be single peculiar species with wide ranges. By contrast, the evolution of a generalised, versatile body build tends to be multiplied into a great many minor variations on a single theme. Antelopes, kangaroos, finches, pigeons – the list goes on – all have many closely related species and subspecies.

Some of these radiations keep that diversity going for millions of years. Others have a brief flourish and are overtaken by a superior relative. For example, fossils betray that many types of relatively sluggish rodents used to be both diverse and abundant in Africa before they were overtaken by the fast and omnivorous rats and mice.

The first savanna apes may have enjoyed such a flourish. In the early days of a successful ecological shift, such as when apes moved into the savanna, it is likely that there would have been a rapid diversification. Like modern African monkeys, those apes would have formed distinct regional populations, larger and smaller types, runners and skulkers, rarer and commoner ones. The known fossils show some of this variety; notably pygmies and giants, slim omnivores and mill-toothed herbivores.

If the immediate predecessors of these hominids were savanna apes that had dispersed from the Cape of Good Hope to the Horn of Africa, can we match up such specialisations with the regions that were outlined on p. 19? For fundamentally tropical founders of the hominid lineage, East Africa would have remained the area of greatest continuity in all aspects of the environment, even during major climatic perturbations. For hominids, the exceptional physical diversity of eastern Africa should have favoured maximum flexibility and adaptability in diet and behaviour. Because lucies (*A. afarensis*) were the most generalised form of australopithecine I suspect that they were the most likely to have had specifically eastern African origins (although they are known to have reached all four regions).

By contrast, the southern and northern extremities would have had at least one major feature in common to distinguish these areas from tropical East Africa – cooler temperatures. Cold tends to limit the number of species including prey, competitors, parasites and diseases. This in turn could have altered the patterns of density, dispersal and survivorship in hominids. Over thousands of years all these factors would have helped generate the genetic differences on which evolution could work.

In the extreme south late summer fires lit by lightning would have been a natural attribute of the annual cycle. As a result, endemic plant communities (and many of the smaller animals) were very well adapted to cold, fire and drought. This means a higher preponderance of plants that were difficult to chew, process or digest. Any reliance on vegetarianism in the first hominids would have faced, literally, tougher times in the south and stronger pressures to evolve strong teeth and jaws.

Given these tendencies I suspect that precursors of the Nut-crackers were more likely to have evolved from a southern rather than an equatorial population of savanna apes.

The possibility that fossil-hunters have so far failed to unearth a southern ape from the human main-line lineage gets some support from the first fossils that are unarguably *Homo,* beginning some 2.4 million years ago. Once again, the earliest fossils of *Homo* show great variation. They have many resemblances to some of the australopithecines (predictably to the earlier, least differentiated ones), but there is no simple straight line. Both *Australopithecus* and early *Homo* formed highly variable clusters, and the members of each genus have a mosaic of peculiarities. This should surprise no one if these animals conformed to the normal pattern of regional differentiation and minor specialisation. They were not constrained by a very narrow niche or a unique food-getting technique, so what was there to stop diversity? The pattern can be likened to a pagoda tree with a succession of branchings following one another. The main difference is that there is no preordained central trunk and any branch might provide the start of a new radiation. The currently known time span of the australopithecines runs from about 4.1 to 1.2 million years BP (before the present). Its real span may have been much longer with many more local forms, but long before this lineage died out, a more advanced *Homo* cluster began to become dominant. Nonetheless, the early ones show quite clearly that they share some characteristics with their predecessors.

In Olduvai Gorge, the tiny skeleton of a female hominid, who would have stood about 106 cm (42 in) tall, is a mix of australopithecine and human features. In the absence of a clear identification she is known as Hominid 62. By contrast, another fossil that combines southern ape and human traits is of massive size. The two, three or more early human types from this period have been lumped together under the jokey title of 'Handyman' or *Homo habilis.* They represent the second tier of the pagoda tree.

The teeth of one later form of 'habiline' have proportions that resemble those of Modern humans. This form has been variously allocated to *habilis,* to early *erectus* or to a species in its own right – 'action-man' or *H. ergaster.* Should this fossil prove to have the most direct links with Modern humans, then *erectus* could prove to be a late offshoot from our own lineage. It is wiser to regard fossils as guides for a known progression than to appoint them as actual ancestors.

Ancestors or not, the fossils are real enough and they certainly

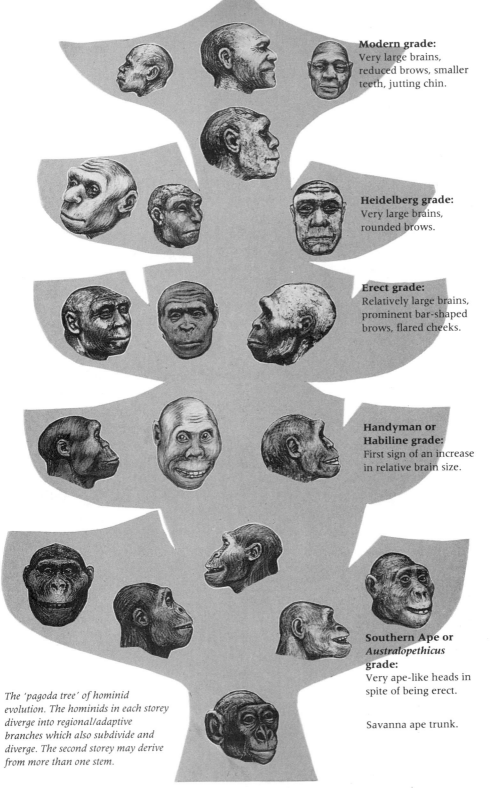

Modern grade:
Very large brains, reduced brows, smaller teeth, jutting chin.

Heidelberg grade:
Very large brains, rounded brows.

Erect grade:
Relatively large brains, prominent bar-shaped brows, flared cheeks.

Handyman or Habiline grade:
First sign of an increase in relative brain size.

Southern Ape or *Australopethicus* grade:
Very ape-like heads in spite of being erect.

Savanna ape trunk.

The 'pagoda tree' of hominid evolution. The hominids in each storey diverge into regional/adaptive branches which also subdivide and diverge. The second storey may derive from more than one stem.

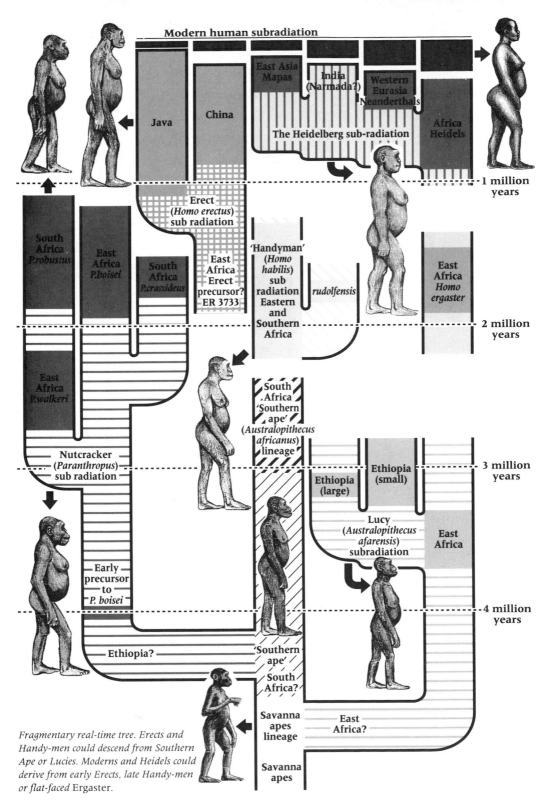

Modern human subradiation

East Asia Mapas

India (Narmada?)

Western Eurasia Neanderthals

Java

China

The Heidelberg sub-radiation

Africa Heidels

1 million years

Erect (*Homo erectus*) sub radiation

East Africa Erect precursor? ER 3733

'Handyman' (*Homo habilis*) sub radiation Eastern and Southern Africa

rudolfensis

East Africa *Homo ergaster*

South Africa *P.robustus*

East Africa *P.boisei*

South Africa *P.crassideus*

2 million years

East Africa *P.walkeri*

South Africa 'Southern ape' (*Australopithecus africanus*) lineage

Ethiopia (small)

3 million years

Nutcracker (*Paranthropus*) sub radiation

Ethiopia (large)

Lucy (*Australopithecus afarensis*) subradiation

East Africa

Early precursor to *P. boisei*

4 million years

Ethiopia?

'Southern ape'

South Africa?

Savanna apes lineage

East Africa?

Fragmentary real-time tree. Erects and Handy-men could descend from Southern Ape or Lucies. Moderns and Heidels could derive from early Erects, late Handy-men or flat-faced Ergaster.

Savanna apes

35

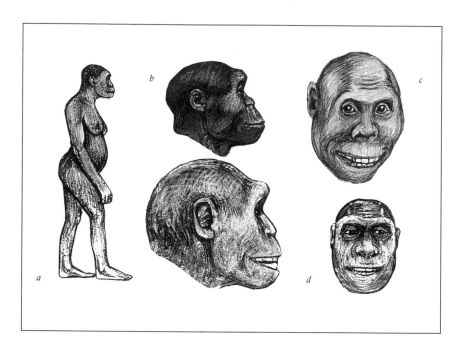

illustrate the sort of developments that were taking place between 2.3 and 1.6 million years ago. The most striking aspect of these is increase in *relative* brain-size or encephalisation quotient (see above). The teeth had shallower roots and were aligned along a parabolic arch. The lower limbs became longer, while the upper arms became shorter. The thumb became larger and was fully and strongly opposable. If that oft-repeated diagnosis sounds flat and uninteresting it is because it is shorn of the broader adaptive changes that give functional forms their ecological meaning and context.

The anatomical innovations of emergent *Homo* find their meaning less through these descriptive details than through their association with one momentous fact. An African primate had crossed an entirely new threshold where biological adaptation was overtaken by cultural adaptation. Anatomy continued to be a meaningful manifestation of change, but it became both subordinate to and, to a significant degree, the *consequence* of culture. The enlargement of brains that distinguishes *Homo* from *Australopithecus* coincides with the first signs of a stone tool tradition. It is tempting to see a correlation but the connection is not a direct one.

The invention of a tool has less implication for brain size than the ability to impart knowledge of its effective use. A chain of learning may mimic inheritance as the pathway for a package of information but the survival of any technique within a human culture is hazardous. Geneti-

The Handyman 'grade' of human evolution illustrated by reconstruction of Homo habilis *(a) standing and (b) in profile. (c) The 1470 hominid. (d) Profile and frontal view of the flat-faced 'action-man' or* Homo ergaster.

cally programmed skills such as an oystercatcher's catching or the hatcheting of a nuthatch are superior to artificially invented technologies in that almost all the skill components, from identifying the correct context for the 'tool' to its actual use, are built-in and heritable. By contrast, a human's inventions die with him unless he can pass his skills on. Given the hazards of human existence and the number of skills that may have to be learnt, we should look to mental quickness, adaptability and aptitude in the teaching and learning of new skills for the greatest survival value.

Skills in the teaching and learning of tool-use imply an increase in the capacity (and therefore the volume) of the brain because Stone Age education, no less than any other, must have been a process and a programme. Programmes allow both the execution and learning of more complex techniques to proceed from a baseline of simpler ones. It is not necessary to posit language for the earliest human polytechnics, nonetheless the charades that would have been needed to communicate must have had the germs of a gestural language in them.

In an extraordinary mimicry of natural adaptations, these primates began to manipulate elements and use materials in a way that rapidly multiplied the number of ecological niches they could invade. Each new tool opened possibilities that were formerly the prerogative of very specialised animals. Where diggers had needed heavy nails, now there were stone picks, cats no longer had the monopoly of sharp claws, spears mimicked horns, porcupine quills or canine teeth and so on. Here, for the first time, was an animal that was learning a multiplicity of roles via the invention of technology. An increasing number of animals now had a new competitor that would encroach on at least a part of their former niche. In some cases (perhaps some of the scavengers) the overlap may have been so great that the hominids took over.

Accumulations of stone tools and other debris have evoked a variety of interpretations which in turn have invited prehistorians to take up different theoretical stances. It is claimed that early humans only ate alone or they shared food, they were wholly nomadic or they fell back on home bases, they cached tools and carried food to them or they cached food and carried tools, they butchered large animals or they were merely peripheral scavengers, meat was a rare or a frequent item in their diet. There is persuasive evidence for each of these positions but most models for the foraging patterns of early humans have tended to suppress awareness of the ecological diversity of African ecosystems and underplay the behavioural plasticity of primates, let alone an advanced human one.

A picture of the Handyman's handiwork emerges from the lower beds in Olduvai. There is evidence here that these hominids butchered animals and transported both carcasses and stone tools (the latter from as far away as 11 km). In a detailed analysis of prey species, body parts and the various cuts, crunches and abrasions on bones, the taphonomist Richard Potts found some bones where tool cuts overlay tooth marks – obvious evidence for hominid scavenging.

He also found the reverse sequence and a notably high proportion of butchered forelimbs taken from medium-sized bovids. Since nonhuman scavengers are better equipped to detect carcasses more quickly Potts' inference that the Handymen hunted these animals is reasonable. In spite of this, many other anthropologists have been unable to envisage such early hominids having the skills, organisation or tools to trap, ambush, catch or disable prey. However, it is almost inconceivable that these omnivorous hominids failed to take advantage of the numerous and vulnerable prey that became available during seasonal gluts. Many birds and mammals of all sizes concentrate, migrate or go through booms and busts. It is very likely that skills in trapping and hunting were both tested and improved at such times.

Skeletons of the Handyman are less well known than those of the Australopithecines and of the Erect humans that came later in the Pleistocene. However, the human erect posture, versatile hands and humanoid foot would all have been refined during this evolutionary phase.

One way of portraying anatomical changes from this point on is to sketch in a sequence, whereby technical versatility so increased that the momentum of cultural change *itself* might have helped to alter and transform the physical shape of humans.

That expansion had several interconnected parts. First came invention, sometimes slow but, nonetheless, a continuous expansion in technical virtuosity and scope. This enlarged the range of habitable environments and of foods to be eaten, and therefore allowed more humans to exist at higher densities over a wider area. This in turn could have involved becoming more manipulative, more flexible, more playfully experimental. While the amount of knowledge and skills an individual needed would have benefited from a longer and longer learning period, human technological invention made the playful nature of childhood a functional virtue in its own right. The infantile shape of modern skulls is, in a sense, the anatomical expression of multiple skills in communication, manipulation, language and coordination.

The paedomorphic or infantile trend that marks out human evolution hints at links between anatomy, the development of technology and cooperative behaviour. Furthermore, children themselves are prime targets for natural selection. In this the ultimate agency may be external diseases, predators or accidents but a more immediate selective force is exerted by other members of its own group or family. If, for example, there is a potential for parental neglect there are advantages for the young that 'manage' their parents best as well as for those with a responsible mother or father.

To illustrate the control exerted by young animals over adults, consider how the newly born or hatched summon their parents with a cry. That wail or shriek is much more than a simple expression of need, it is the young animal's only means of forcing its elders to give it attention, protection or food. Its coercive effect probably has evolutionary roots in the danger, to family or group, of alerting predators. Such cries are soon augmented, developed or differentiated as the growing youngster's range of needs widens. Both cry and begging behaviour may change continuously; eventually it may even get incorporated into adult situations where fellows have to be appeased, cajoled, seduced or courted.

One such infantile behaviour that is common to all higher primates is the tantrum but when the contexts and character of chimpanzee and human tantrums are compared there are significant differences. Where the young chimp's pique is a response to another chimp's failure to allow the infant access to food or fun, the human child goes further than 'Attend to me or I'll scream and stamp!' The human shouts, gestures or pulls to assert a wider range of messages – 'Look at me, see my work – I'm smart.' The context for these efforts to force attention out of adults is often linked with learning and more specifically with the acquisition of technical skills.

Children coerce parents by screaming. Negrito child, Baclai, Philippines (Photo J. Kamminga).

The more skills or 'niches' that humans have appropriated, the more versatility and flexibility has to be learnt. This is in marked contrast to any other animal. All the lower animals are rigidly programmed to perform their genetically predetermined roles. More developed species augment this with behaviour learnt by imitation of the adults. In apes, the youngster's main incentive for imitating its mother seems to be sharing her food and keeping close to her. By contrast, a prime incentive for a human display of imitative skill seems to be parental praise (something that is rewarding to both child and parent).

Because so many technical skills have to be learnt, children are constantly competing among themselves or for their elders' attention. When this social dimension of learning is projected back to the earliest forms of *Homo* the need to teach simple techniques to children stands out as one of the most significant forces that might have shaped human evolution. Indeed, it may be in the biological origins of our earliest education techniques that humans can most truly be said to be self-made.

There are immediate practical benefits for the adaptable learner of new techniques but there is a less obvious evolutionary advantage for those children best able to display their skills in different roles. It is here that the showing-off or tantrum element may be most significant. Anyone familiar with a school yard will know the many faces of juvenile competition. Here and in the family admiration and approval reward the rapidity and conviction with which a child plays its various tasks or roles. The younger the child, the more play is the operative word. Although its task may contribute something to the daily life of a foraging family it is not sharply demarcated from other expressions of juvenile energy such as clowning or mock fighting. Nonetheless, each 'play' is one act in an ever-expanding repertoire. Specifically adult activities are not the only models for mimicry; the noises, gaits or displays of other animals and other children are part of the store of imitative performances.

Young apes like to strut around, stick flowers or leaves on their heads and generally show off to one another and to adults. The most conspicuous difference between the high spirits of ape and human children is the latter's flexibility in asserting 'self' in a variety of roles. I believe that the progressive development of self-consciousness, the awareness of role-playing and the ability to express that awareness is much more likely to be the product of competition and selection among juveniles than at the adult stage. Smart kids able to manipulate others in their own favour had the best chance of surviving. The tying in of skills with artefacts and social relationships therefore socialised the learning of technological skill in a unique way.

Because *Homo* is an animal that has specialised in slowing down growth, the presence of children permeates human society as no other. Childhood is not only a longer proportion of a lifetime but children's needs have to be accommodated into many if not most adult activities. Prolonged involvement of *Homo* with their children would have eased the diffusion of childish innovation into adult practice.

Infantile role-playing charades that use symbolic gestures or sounds contain the rudiments of language. Once codified to suit adult purposes the signals invented during play can easily help to coordinate and control actions. A diversity of signals would have introduced the elements of choice and direction to decisions (regardless of how these were reached). Having a knowledge-sharing code that allowed all members of the group to benefit from the short-term forays or longer-term experiences of single individuals would have had many selective advantages. For example, it would have allowed a group to hurry to a food site faster than other less flexible animals. In this may have lain the seeds of a uniquely human culture where language became the medium orchestrating all group activities.

Older experienced animals are critical for the maintenance and diffusion of traditions in many species, but I think it unlikely that the more mature hominids were sources of innovation, as is often assumed. It is more likely that inventiveness would have come from the younger and less experienced in playful but nonetheless ecologically relevant contexts.

There have been innumerable attempts to draw a boundary between animals and humans and identify the elements that nurtured our humanity. I am confident that when scientists eventually seek to learn the vocabulary and architecture of ape communication systems they will find tiny hints of our own articulate self-awareness even if it is couched in some muted form. Here they will find a smudged but decisive transition. Just as the development of a foetus imitates an animal's earliest evolution, the born child's rapid passage through a sequence of changes is another form of reliving evolution. Whether we seek to understand how self-awareness emerges in a living individual or look for its emergence in extinct populations of hominids there is much to be learned from a comparison of the psychological development of infant humans and chimpanzees.

Of course the socialisation of an individual child cannot recapitulate evolution, but it *is* a process whereby a smaller number of relatively simple components become transformed into very large complexes by observable steps. If the main characteristic of our lineage has been to extend childhood, then analysis of early stages in the development of an individual should be a real guide to our intrinsic nature. I call this evolutionary dimension, this concealed entity, the unknown child. It is a condition every one of us has been through.

The third tier of the pagoda tree, the *erectus* radiation, embraces a long-lasting and far-flung assortment of fossils.

The name *Homo erectus*, the Erect man, was first used in 1895 to describe a single skull cap found in Java. Since then, *Homo erectus* has come to accommodate a large number of fossils which date from about 1.6 to 0.1 million years ago. These include many African fossils that go back much further than the Asian ones. In addition to being earlier, the

The Erect 'grade' of human evolution illustrated by reconstructions of Homo erectus pekinensis *(standing and in ¾ view (after a Chinese model). Sketches of reconstructed* Homo erectus *from Solo, Java on right.*

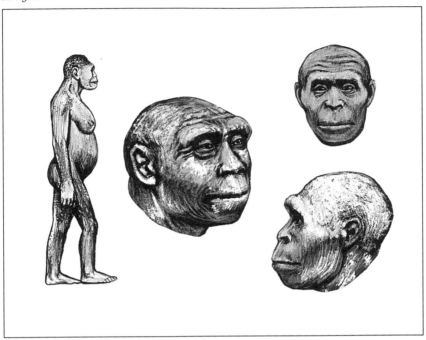

African specimens are hard to differentiate from late members of the *habilis* cluster.

Although the two Asiatic populations (one in China and the other in Java) clearly share a common ancestry with the earliest Africans, there is no doubt that the Asians developed off at a tangent, acquiring many local specialisations. The African subspecies is called *H. erectus olduvaiensis*, the Javan *H. erectus erectus* and the Chinese *H. erectus pekinensis*. This advanced human, the first to leave Africa, diverged into an unknown number of distinct populations. If and when Indian specimens are found it is quite probable that other subspecies will have to be named.

Erect men showed some differences in both the lower and upper limbs and even subtler ones in the pelvis, but, to all appearances, their body-build resembled that of Modern humans very closely. Generally larger than *habilis* types, they had proportionately thicker crania and their bones were much denser and more massive than those of Moderns. In terms of proportions the Erect human brain made little appreciable advance on the *habilis* types.

The Heidelberg 'grade' of human evolution illustrated by reconstructions of Broken Hill (standing, profile and frontal). Neanderthal from Shanidar, Iraq (centre), Neanderthal from La Chapelle, France (right above), and Mapa from China, after a Chinese model (bottom right).

With such diversification spread over 1.5 million years and such a vast range, where do Modern humans fit in? There is a school of thought that regards Erect humans as the direct precursors of modern people. For them the family tree is short and very top-heavy. But anthropologists now look increasingly towards a fourth tier of the pagoda tree for the emergence of our own ancestral branch.

This fourth cluster of prehistoric people is still very imperfectly understood. Once again, the great regional and individual variety of all hominid types has led to many names, and for a while they were given the confusing and contradictory appellation of 'archaic moderns'. A leading primate taxonomist, Colin Groves, has resolved the difficulties by calling the African and European ones Heidelbergs (*Homo sapiens heidelbergensis*) and the Asiatic offshoot Mapas (*Homo sapiens mapaensis*). Many specimens now regarded as belonging to these two types used to be described as 'late *erectus*', such has been the state of confusion as more and more human fossils have been unearthed. Heidelbergs and Mapas manifest one more cluster that proliferated and diversified as they too spread across a large part of the Old World.

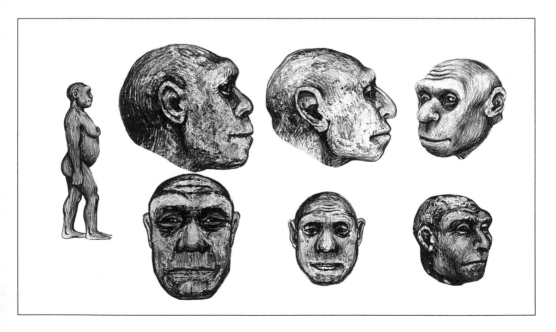

Fully Modern humans (the fifth tier of the pagoda tree) are the concern of subsequent chapters and it is enough to say here that they appear to have differentiated from a Heidelberg population in eastern Africa. The place and time of Heidelberg emergence remains one of the least understood phases of human evolution and, while either an African or Eurasian origin is possible, the former is the more likely. One rationale for elevating these superficially *erectus*-like humans to the status of *Homo sapiens* was an increase in brain volume (concealed behind low foreheads and prominent brows). There were also other details of skull structure that would ally them more closely to us than to *erectus*. For example, their brow ridges follow the curve of the orbit and thin out to the sides, whereas Erect brows formed a solid bar that flared out to the cheeks. Heidelberg heads were generally less bony and angular than those of Erects. They belong to a confusing but highly significant stage of human development, one which must await more fossils before it can be adequately interpreted.

In apes and all early humans the eye sockets, nose and teeth are tied up together in a single unit. This unit is cantilevered forward off the cranium over a series of struts and bridges which enclose 'the face' within an outer scaffolding or framework. This facial unit has a lower boundary in the strongly reinforced arch of the mandible, while its upper boundary is the brow ridge, also strongly reinforced. This reinforcement is especially marked over the deep roots of ape canines, but it can be said that all other activities of the head are essentially subsidiary to those of biting and chewing.

A simple way of summarising the changes between ape and Modern humans is to examine the components of the head as plastic mutable units that have responded to evolutionary change. The most dramatic of these changes concerns the role of the teeth. The canine and incisor teeth have been most transformed. Instead of having long roots and thick, deeply implanted canine corner posts, the human tooth row is feebly rooted and its principal strength now lies in an uninterrupted arch of similar-sized teeth acting in concert. Having lost their anchoring role and stabbing function, human canines scarcely punctuate the smooth transition from weakly cutting incisors to stronger chewing molars. In the course of becoming shallow-rooted and being served by ever-smaller chewing muscles, the shrinking tooth row shifted backwards and up closer to the larger cranium. The development that gave us an intrinsically human face rather than an apelike muzzle is partly due to a simple switch-over in where the face was anchored.

The human mandible has retained its strength and basic form as a structural component of the face but the reorganisation of both tooth rows into a shallow arch has taken place relatively independently of the surrounding bonework. One consequence of the lower arch of the teeth shifting back at a faster rate than the parent jawbone is that the chin juts out strongly from a tooth row that has retreated further back.

Shifting the tooth arches back has had an even more dramatic consequence for the upper boundary of the face. Declining muzzle and enlarging brain converge upon one another and there comes a point when the brow ridges become redundant, because the ballooning forehead comes to provide a broad and continuous anchoring surface in place of the projecting ridges that had served to protect the eyes, frame the face and anchor the tooth rows.

An innovation that has set Moderns apart from other early humans is a dramatic thinning and lightening of all our bones, including the cranium. This lack of massive bone resembles 'arrested development': it is as though we retain a juvenile lightweight skull and skeleton. Compared to skulls of large apes or extinct types of men, our heads are decidedly childlike and this is not simply a matter of face proportions. Maturation in a Modern human transforms the adult face less completely than in an ape or extinct man, because we fail to thicken up our bones with the massive ridges and buttresses of a truly robust animal. Like a soft orange ripening within its thin rind, the globe of a growing child's brain forces the capsule of plastic bone that surrounds it to accommodate to its steady enlargement. The exact proportions of an infant's skull can be easily altered by swaddling (which makes for short, broad skulls) or deliberate distortions through binding the head. Artificial shaping is made all the easier by the typical delicateness of a Modern human's skeleton and the fact that the individual bones of the cranium are soft and disconnected in a newborn child.

Human hands, though only slightly modified from those of other primates, have been the organs that permitted changes to take place. Their ability to fashion tools and contrivances enabled humans to buck a universal rule that governs the evolution of other animals. They could

Oldowan pebble tools and handaxes were made by controlled striking of stone against stone.

escape all the modifications to anatomy, physiology and behaviour that go with becoming a dietary specialist. The first tools helped augment the natural anatomical versatility of primate arms, palms, fingers and fingernails. Apes can use levers, probes, sponges, hammers and flails. The earliest hominids not only used hammer stones like modern apes do but, about 2.5 MY (million years) ago, learnt that pebbles could be cracked against one another to generate cutting edges. For an animal with a strong preference for river courses, pebbles were the most freely available hard material that could be found. Naturally occurring 'blades' would have to be searched for, whereas pebble tools could be made from a very abundant resource.

Because pebbles were common, heavy and very easily modified, the tools were simply made, used on the spot and abandoned. We can be certain that this opportunistic tradition would have come to include the modification and momentary use of many other materials, such as pieces of bone, horn, sinew, skin, shell, wood, bark, fibres, lianas and grass. In confirmation that pre-Modern humans used stone tools to shape wood, a small wooden 'plank' has been excavated in the Jordan valley. Found

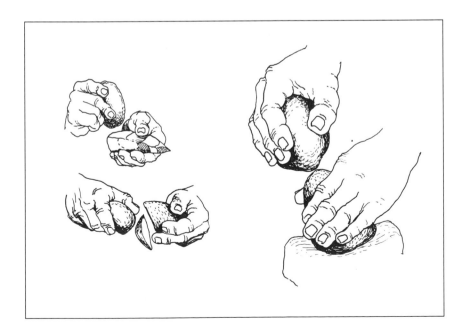

close to a layer with many stone hand axes, the wood was flat, cut across the grain and then polished. Although it is relatively late at 750,000 to 240,000 BP it could be the handiwork of either Erects or Heidelbergs.

The crushing and cutting of tough foods would have represented a small part of the overall activity of early hominids. If archaeology has given pebble tools so much importance it is only because they are the sole surviving evidence of early human activity. Nonetheless, it is certain that hominid bipeds were faced with numerous problems to which they brought intelligent observation, deduction, strong hands and the ability to carry and assemble materials. The behaviour of living apes implies this much, and excavations have shown that the first hominids cluttered their resting places with more than pebble tools and animal bones. For example, it is possible that the 'Handyman' was using wooden structures in at least one African site.

Evidence for the building of an artificial shelter has been claimed for an occupation floor in Olduvai Gorge dated at about 1.8 million years ago. A stone circle, interpreted by its excavators as the footings for a branch shelter or hut and measuring about four metres across, was surrounded by an extensive scatter of tools – choppers, scrapers, anvils and hammer stones – as well as artificially introduced stones of no obvious function. Broken bones of elephants, rhinos, hippos, giraffes and many species of antelopes, zebra, pigs as well as numerous other small animals litter the site.

Although some archaeologists dispute the evidence for a hut this living area does suggest that the carriage of heavy artefacts, bones and pieces of animals was well developed before the *Homo erectus* stage of evolution. An early use of some form of carrier bag is implied by large numbers of stone tools that occur far from their original outcroppings in this and in some other later sites. It has been suggested that some rock tool piles were caches to which carcasses were dragged for processing. Wherever and whenever shelters or windbreaks first began it is very unlikely that branches would be simply stuck in the ground without some form of interweaving or tying, and the use of bush string, natural rope or cord was as important as the use of stone tools in the evolution of human cultures.

Chimpanzees interlace branches for their nests, and many birds thread, weave, sew and knot thin fibres while nest-building; they also choose three or four different classes of materials in succession. Many creepers have abundant fruits which are more easily harvested by pulling down and stripping the vine stems. These lengths of vine, like strips of skin, sinew and gut, make excellent material for tying. An observant hominid would have noticed numerous examples of animals detained by a variety of natural tangles or webs. Large spiders can catch prey as large as small birds. It is not uncommon for birds, up to the size of a young cassowary or eagle, to become entangled in bushes or canes with barbed or sticky stems or foliage. Likewise, weak animals sometimes get their neck or legs snagged in thickets with trailing vines and thus become strangled or held fast. At some rather early time in human evolution observing and benefiting from such mishaps would have been augmented by the contrivance of their replication by means of crude snares. Large animals commonly get bogged down in the wet season and further entanglement of animals already waylaid by mud would have been easily assisted by manipulating trailing vines and branches.

Early hominids would have followed animal paths as a matter of course (indeed most travel would have depended on them). They would have noted trouble spots for the unwary, would have learnt where and when each species preferred to travel and could have intercepted or diverted prey into prepared ambush or trap sites. Both the incentive and the skills to invent this sort of contrivance would have long predated the emergence of Erect humans. They would have begun with the use of natural cords and vines *in situ* but, because vines could have helped with the porterage or dragging of food, tools, stones and branches back to the camp, the use of cordage within the living area would eventually have caught on. A strong 'feeling' for string, cordage and basketry is evident in all gathering, hunting and fishing communities up to the present. Men and women spend long hours processing, repairing and tying fibres into nets or weaving ingenious traps. Games are played with 'cats' cradles' and the breaking points of various fibres are not only discussed and well known but also tested in tug-of-war games. Some of these traditions probably have a history of 2 million years. In my view traps are likely to have harvested much more food than active hunting did throughout most periods of prehistory.

The great majority of prehistoric artefacts must have been constructed from animal or plant material. Since these have not survived, how can

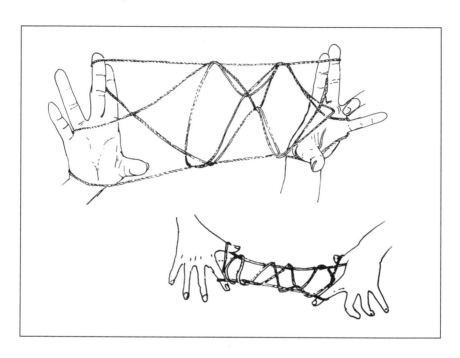

we retrieve any understanding of that immense lost industry? How can we begin to inventory their diversity and appreciate their fitness in terms of functional form? How tease out their possible origins?

Perhaps a start can be made by recognising that the material for artefacts and the main commodities of prehistoric existence spring suddenly into use through the very act of predation and the decision to collect or pick up an organism. To a significant extent the use of organic materials and artefacts is subsidiary to plain appetite.

At the moment when a prey animal is killed its independent otherness departs and it becomes a possession, a mass of diverse materials, which must in the literal sense be rapidly sorted out by an owner or owners before they are spoiled. Typically, the animal's outer membrane, its skin, is transformed from an essential biological entity into a material that its new owner can either use or discard as debris. One potential for skin and its most immediate usefulness is as the carrier-bag in which the whole or parts of its contents will be transported from the kill site to the eating, sharing or cooking site. Thereafter it may be processed further and find different uses or it may be discarded, but that brief utility of the skin as a carrier-bag illustrates some significant characteristics of the earliest artefacts.

One is that the material is right there on the spot, just where there is a need for it. Anyone who has forgotten their bag or haversack will know how awkward it is to carry more than one or two small items, let alone the various and often large masses that prehistoric people would have had to carry. The other demand is for material economy; weight and bulk must be at a minimum when there is a long walk back to the group or to a base. Another consideration is multiple use – if an animal skin is found to be of further use, say as mats, body covering or thongs, choice of that material over alternatives may provide an added (or even prime) incentive for hunting a particular species in the first place.

Part of this choice involves quality. Many subsistence hunters concentrate on certain prey species for the quality of the by-products – hides that have particular properties such as flexibility, softness, elasticity and impenetrability will be preferred to ones that tear easily or rot quickly. The warm fur of an otherwise inferior prey may make it more desirable.

Economy of effort is another prime consideration, so both the pliant accommodation of fresh skin to its load and the ease with which it can be separated from tissues commend it. On the other hand, its use is dependent upon the hunter having a sharp blade. Any material will be useless if the means of transforming it to specific purposes is not immediately at hand. A single precondition, such as a blade, may be attached to still further preconditions.

For example, the earliest hunting hominids probably opened up large mammals with slivers of stone, bone, tooth, shell or plants which could be improvised from materials in the immediate vicinity. However, in swamps or plains, where few or none of these materials were at hand, hunting would have been severely constrained unless blades were carried in preparation for a successful hunt. Among contemporary hunters a neat solution is to carry a detachable multipurpose bladed spearhead that can kill, butcher, skin, trim and whittle (but, of course, the earliest spears were probably little more than sharply pointed poles).

Since hands must be free for most routine activities, including the hunt, taking along a blade would have required not only a tie or thong but the blade itself would have required some sort of envelope. This shows that the most elementary processing of a primary product, such as the hide of a mammal, not only implies but effectively *demands* the use or development of further technology involving further working of materials. This dynamic helps explain how technologies are elaborated.

Plants also illustrate how artefacts can emerge in the course of satisfying appetite. For early humans invading new territory or habitat, one of the surest guides to both the edibility and the whereabouts of plant foods would have been the feeding habits of other animals. Most animals learn to eat from watching or being fed by their parents and, at the level of a single life, innovation and experiment in new foods is not usually significant.

We tend to assume that animals and 'natural people' know what is good for them, but huge numbers of domestic stock die every year from eating poisonous plants and there is evidence of dietary disease in prehistoric man. One of the most interesting of these concerns a tall youth of *Homo erectus*, whose skeleton was studied by Alan Walker at the Johns Hopkins Medical School. Walker found a condition called hyper-vitaminosis that is associated with overconsumption of vitamin A, which is concentrated to very high levels in the livers of carnivores. He suggested that *H. erectus* around Lake Turkana a million years ago could have been preying upon some of the competing local predators. Among the most likely victims would have been hyaenas, jackals and perhaps crocodiles.

At the broader level, and in the context of whole tribes of people expanding into new habitats and climatic regions, each dispersion must have involved learning and dietary experimentation as a direct by-product of territorial expansion.

For example, colonisation of the moist tropics by Erect humans would have involved many important changes in diet. The ways in which new foods could have been discovered can be listed as follows: 1) observing animals feeding; 2) famine-induced experiment; 3) curiosity-induced experiment; 4) by-products of other activities; and 5) chance accident. For *Homo erectus* the most important source is likely to have been the observation of other animals and the sampling of their left-overs.

Large groups of primates, elephants and giant pigs all do great damage to plants in the course of feeding. (In most instances the plants have evolved to endure this, perhaps for the benefit of wide dispersal.) For prehistoric humans this massive impact of animals on plants would have been a routine experience that offered the opportunity of revealing food sources otherwise hidden underground, out of reach, or broken open by strength no human could emulate. Palm pith, many edible roots and numerous fruit foods would have owed their first discovery to simple scavenging in the wake of elephants, rhinos or giant pigs. Not only the food sources themselves would have been significant, but the context of unearthing, breaking and opening would have provided a ready-made abundance of struts and levers to assist effective scavenging. More earth would need to be moved to get at bits left behind, broken branches shaken or wielded to get at pockets of fruit, and slivers of palm would lie about to dig out more pith.

Whether known or not, the object's material usefulness and its properties as a food or fabric become open to assessment. Hunters constantly compare: this young animal will be tender, that old one will be tough; this species has fat of a fine flavour, that one less so. At one season a mollusc or plant will be tasty, before too long it may become inedible or even poisonous, in between fine judgements may be necessary.

Comparison or assessment of condition, quality, flexibility, age and so on is not an exclusively human skill. Parrots will gently squeeze and tongue-probe fruits for their sweetness before plucking or passing on; carnivores visibly scan a herd before going for one particular straggler or limper. I have seen a heavy male chimpanzee momentarily stress a slender vertical sapling between his fists before trusting his whole weight to a rapid ascent. By keeping the stem rigid between three limbs, he was able to climb ten metres on a support that would have snapped off at once had he lost his balance or failed to maintain rigidity in the section he was travelling up. Such a performance demonstrates a finely tuned awareness of the properties of a plant that was usefully put to the test by that chimp. We should at least credit early humans with a comparable or superior ability to use certain properties of animals and plants to their own ends.

Similar use of organic materials (in an opportunistic or incidental way, but also on a more systematic basis) must be assumed. It is obtuse to deny that early man possessed simple skills because they cannot be proved. This is to prune his diet down to the bones in his middens, to impoverish his versatility by disallowing all but a handful of stone implements. Humans living in the tropics occupied rich and diverse environments with a known variety of foods – not 'potential' foods. There are lengthy lists of accessible and edible fruits that would have been widespread and common within their own ranges. Likewise, there are entire communities of animals that would have been well suited to being harvested without involving difficult processing or highly developed tools.

These resources were used by Modern and recent peoples up to the present or very recently, but almost all were much more abundant and concentrated in the past. Turtles, migrating birds or herbivores, floodwater fish and insect swarms are obvious examples. A positive way of trying to understand the economies of prehistoric peoples is to examine the fluctuating largesse of tropical ecosystems as they exist today (even in vestigial form) and seek the particular problems they would have posed for those trying to exploit them.

Study of the behaviour of a succession of prey animals, together with their seasonal cycles, can build up a calendar of events that would have been significant for prehistoric people. Such studies give an idea of when and where these animals would have been easiest to harvest (for example, by locating the strategic vulnerability of certain fish in seasonal or tidal shallows). The physical *means* of harvesting such resources (such as using fences to corral fish behind retreating water) can often be inferred from *in situ* plants that are easily modified or manipulated to serve as fencing. Indeed, these plants might have suggested the strategy in the first place by causing natural barriers – mangroves by the sea, reeds, sedges and fallen palm leaves on inland waters.

Of all the tools available to prehistoric peoples fire is quite the most significant and it is with the widespread use of fire that they began to shape the landscape. Originally fire would have been restricted to certain very limited regions, seasons, habitats and situations; most obviously close to volcanoes. Occasionally, oil, shale, bitumen or mineral oil seepages can be ignited and marsh fires are commonly triggered when methane gas is lit by lightning or by sun rays focused by dew drops above dry tinder. Lightning strikes are probably the most common starter for fires, but they only take hold and spread when both wind and fuel are right. Numerous plants are adapted to withstand regular firing – a capacity that was evolved long before humans came along – but the regions in which those plants evolved were originally quite limited: the Cape of Good Hope, parts of Central Africa, Australia and India for example.

In some fibres and woods embers can smoulder for long periods without erupting into open flames. This is particularly true of underground burns where lightning ignites dead roots, peat or buried forest leaf-fall and then proceeds much like a charcoal kiln. Prehistoric hominids living in regions where such fires were frequent would have noticed the tendency for long-lasting smoulderings to erupt into open flames when exposed and fanned by a strong wind. This is the type of prolonged burning that would have offered them familiarity with and sustained observation of the behaviour of fires. The situations in which they were of practical use would have been more urgent and immediate.

In lightning-prone regions early hominids would have followed other scavengers to natural fires for the pickings they provided. Not only could fleeing animals be caught in more concentrated numbers, but the aftermath would have been richer still. In the embers could be found scorched pythons and adders, tortoises cooked in their shells and innumerable blackened lizards and insects. If a fire is sufficiently hot and widespread it will entrap much larger animals at vulnerable moments of their life cycle. Very old animals cannot move fast enough and newborn fawns are genetically programmed to lie tight. I have witnessed a huge fire sweeping over a herd of birthing antelopes and seen the outcome: more than a dozen elderly cows roasted, unable to move as they laboured over their last calves.

A host of scavengers, pirates, scroungers, call them what you will, have evolved to take advantage of these natural calamities of life. Before our ancestors out-competed them and appropriated all the major sources of free food there were a lot more various now extinct hyaenas and jackals, bears, marsupial carrion eaters, giant lizards, condors and other carrion birds. Like the kills of large carnivores, natural fires have always created an important resource for such animals and the more omnivorous hominids became, the more they would have joined the throng of scavengers at any natural disaster, led by circling flocks of kites, storks, vultures and rollers. In fire-prone regions hominids would have been acquainted with and perhaps benefited from fire for a million years or more before they attempted to control it for their own ends. We can be reasonably sure that throughout this period they appreciated the way fires force vulnerable prey to concentrate and that they relished the ease with which scorched skin and cooked flesh could be torn apart and consumed. But it would not have been 'cooking' that was the first use of fire.

The most likely first intervention by early hominids would have involved no more than furthering the natural utility of the fire to the individuals that were hanging about for the goodies it offered. Clusters of prey sheltered in clumps of bush that the fire had bypassed could be routed by no more than fetching a brand and renewing the blaze. An ability to keep upwind of a fire and maintain close observation of its progress would have had to precede the first act – but it would have been the easiest and most immediate initiative that would enhance a natural and well-used continuum. The act of renewing the fire would have first revealed that fire could be manipulated like a dangerous ally. Any further steps in its control would have depended upon this hands-on familiarity and possibly with its extension by observing that a fire could be held in one spot by being fed with fuel. However, that awareness is not necessary for other uses to be made of natural fires (and without immediate intervention).

Early hominids were social, opportunistic, vulnerable to large carnivores and biting insects and relatively night-blind. Fire therefore had several attractions. It was warming. Flames and smoke were deterrents and gave some security from nocturnal nuisances. Fire gave light and, perhaps as important, held the almost hypnotic attraction of living 'activity' in the long black inaction and emptiness of a moonless night.

Of all these attractions, countering cold, particularly at higher altitudes and latitudes, would have been of the most practical use and would have provided the most immediate incentive to maintain a control of natural fires. If familiarity with fire was scattered and sporadic, it is likely that different hominid populations flirted with the various potentials of fire for many thousands of years before its control and creation became systematic and routine.

The first firm evidence for a hearth comes from the cave of Escale in southeast France, is dated to about 1 million years ago and is presumed to have been made by Erect man. Effective control and setting of fire probably long predates this, but it is certain that cave-dwelling and human colonisation of cold northern latitudes were totally dependent upon previous fire-controlling techniques that had been developed further south. The first signs of people (rather than cave bears and other carnivores) being present in European and Chinese caves coincides with layers of charcoal and burnt bones.

In European caves the hard, factual evidence of prehistoric menus may be best represented in middens, where the proportions of reindeer, bison or horse bones rise or fall according to the temperature of the period being sampled. As in more recent Eskimo (Inuit) economies in Canada and Greenland, there was a total absence of edible plants throughout the long winters and a corresponding dependence on animals belonging to a very few species.

The far north is a cold snow desert; further south there are arid steppes or hot sand deserts and here too humans had to forgo plant foods for much of the year. Hence there is already a bias against plants showing up in prehistoric sites from these regions, yet these are precisely where the best-studied and best-known sites are – because preservation is favoured by low temperatures, drought and nonacidic soils.

For a contemporary hunter-gatherer the year is punctuated by numerous seasonal events, such as laying or hatching in bird colonies, shoals of fish caught in receding swamps, migrating antelope columns or groves of fruiting trees. Less spectacular eruptions also induce a cycle of movement and change that demands special skills and very varied harvesting techniques. In the past, quite sparse populations of humans would have followed a calendar marked out by periodic excesses of food. Most surviving hunter-gatherers experience short seasons of hunger but more sustained famines mean death for many or in some cases they may induce migration.

It is difficult to reconstruct or even imagine the scale and fluctuating levels of abundance of animals and plants that are now scarce or extinct. Passenger pigeons and bison in America, saiga antelopes in Russia, springbok, quaggas and wildebeest in the Cape were described as 'countless' two or three centuries ago and each came and went in massive migrations or eruptions. The processes and climate patterns that govern, say, wildebeest numbers on Serengeti, locust or rat swarming or a grain harvest on the prairies are now fairly well understood. Periodic bonanzas for prehistoric people were a certainty, and our imagination will probably fall short of visualising their likely scale. It is also not easy to imagine how prehistoric communities coped with boom and bust. We can be certain that migration was one response and that famine and disease would have hit even the smallest communities at times, and that almost all habitats were affected by climatic fluctuations.

Climatically stable zones would have been small in area and almost wholly within the tropics or subtropics. These might have been important regions for the continuity and equilibrium of prehistoric societies and their subsistence patterns. Likely areas can be identified on the basis of known biological refugia, and scattered tracts of Africa, India and parts of the Indonesian archipelago might have served as distinct types of prehistoric human refugia. The African savanna supported those animal and plant communities in which early humans found their most stable and prolonged existence.

Humans ranged through all the more open and well-watered country from the coasts of the Mediterranean through the East African highlands to the temperate Cape. Most of these landscapes were dominated by grass or dry scrub, habitats subject to considerable seasonal and regional changes. Sustaining most of early man's prey, grass itself would not have served him for much else. Bedding, hunting screens, temporary shelters perhaps, but we have little hope of finding anything as ephemeral as a grass bed or hut. Among the species that were directly dependent on grass and provided food for Erects would have been the following in a declining scale of size: several species of elephant, grass rhino, more than one hippopotamus species, zebras, two large buffalo species, various antelopes (including many extinct forms), a variety of pigs, various rodents (notably porcupines and cane rats), hares, ostriches, seed-eating birds and a variety of insects (especially grasshoppers and termites). Potential prey that itself preyed on grass-eaters would have included some carnivores (especially the less dangerous jackals, hyaenas and mongooses), pythons, termite-eating aardvarks and insect-taking birds. Many of these animals would be about or catchable only at certain times of the year; for example, seed-eating or migrant birds, like the antelopes, have to follow the seasonal patterns of their food supply. Browsers or folivores would have greatly augmented the list with giraffes, various antelopes, more rhino and elephant types as well as smaller species from caterpillars to hyraxes. Fruit-eating birds, bats and primates were other important strands in the savanna ecosystem.

The most favourable time for most predators is the later dry season or early wet, when the grass is short from trampling or being consumed by grazers (especially termites) or fire. At such times all life is more concentrated and visible and the grass less hampering for quiet directed movement. Indeed, in many moist areas and on flood-plains rank, dense, waterlogged grasslands would have been uninhabitable during the wet season. In drier areas, where grass is interspersed between open woodland and patchy thickets, the growing season is shorter, the grass thinner and more quickly consumed. It is along river courses within these wooded grasslands that most early human sites have been found. Even where rivers were strictly seasonal, so long as there were waterholes, pools and springs, humans would have been just one of a huge community of animals whose lives revolved around their need for water.

This orientation would have involved more than mere water dependence. Here humans could, then as now, intercept animals that sought water. Interception is a widespread strategy in nature and early humans were probably quite systematic about laying ambushes. They are also likely to have been skilled at relieving the smaller or more solitary carnivores of their prey. Of all prehistoric populations Erect humans are the most likely to have developed such techniques and relied on them for a significant part of their diet.

Since Erect humans dominated the scene for well over a million years and had a decisive role in shaping the world that Modern humans inherited, it is unfortunate that so little is known of their economies. There are some vignettes that illustrate how pre-Moderns may have lived but in almost every instance there are no skulls to reveal whether Erects or Heidelbergs were involved. In later sites, however, the latter are more likely. One of the most important of these sites is on the arid Castilian plateau in Spain. Torralba and Ambrona lie three kilometres apart on the floor of a steep-walled valley at the headwaters of the Ebro River. The site has been documented, thanks to the efforts of Clarke Howell, Leslie Freeman and others, who excavated there in the 1960s. It is still one of the best-known butchering sites for pre-Modern hunters.

Although some twenty living surfaces were excavated they were all from the same period, about 400,000 years ago, and the main prey was elephants. At least seventy elephants were butchered, mostly juvenile but some adult and very old ones as well. Although nearly entire skeletons were unearthed, only one small skull has been found on the excavation site and that had been neatly capped to get at the brain. Brains, like marrow, are a delicacy, so it would seem that these huge, heavy skulls were habitually carried off to be processed and perhaps savoured at sites away from the main butchering ground. The possibility that the actual killing site became a temporary camping ground (as happens today at elephant kills in Africa) is suggested by a peculiar encircling arrangement of bones which might have helped support some sort of circular shelter that has since perished. The tip of a tusk, some 1.3 m long, had been artificially sharpened, most probably to assist its being driven into the ground. Other animals killed in substantial numbers at Torralba are rhinoceroses, horses, aurochs and deer. An unusually scattered pattern of charred twigs led Howell to suggest that large animals were panicked into getting mired in the clay swamps of the site by the use of fire. Freeman went further, and suggested that the charred wood might have been torches and the weighty rounded stones found on killing sites used to pound the mired animals to death. He pointed out that the great number of very large animals systematically incapacitated and killed in the same area must have been the work of more than a nuclear family, and would have easily fed several bands amounting to a hundred or more people. He maintained that the scouting, alerting, signalling, driving, killing and subsequent butchering and sharing could have been achieved only with a complex and flexible communication system – presumably vocal and of a sort we would call language. One opponent of these somewhat hypothetical scenarios has taken guesswork one step further by insisting carcasses were found 'unexpectedly', whereupon the finders were 'forced' to produce appropriate tools.

A much more detailed picture of the shelters that these people could build has come from the slopes of a seashore sand dune on the French side of the Mediterranean, where the people killed a similar range of prey animals but augmented them with shellfish. Dated to much the same period as Torralba and only a few hundred kilometres away, Terra Amata consisted of the remains of several oval huts, measuring between 3 by 7 and 5 by 17 m. Situated near the mouth of a small river on what is now the bay of Nice, post holes, hearths and aligned stones marked out the huts, which were made with posts that probably supported dense leaning walls of branches. Large and small rocks supported the posts and protected the hearth fires from winds. Such large huts could have sheltered up to twenty people. Among the remains are a wooden bowl, red ochre trimmed into pencil-like points and imported flat-faced limestone blocks that may have served as seats.

In the Spanish sites of this period, stone cleavers, hand axes, flakes, burins, scrapers, points and notched blades have been found as well as bone and ivory implements (some probably used for refined percussion on the stones). Spatulas, scoops, blades and scrapers are there, preserved in the boggy ground, and, most significant of all, pieces of worked wood with hollowings, hackings, parings and polishings made by stone and ivory tools – some of the pieces are likely to be the remains of spears.

Another illustration of how intensively the land might have been used comes from the Hunsgi Valley in Karnataka, India. Acheulean tools (which have been assumed to have been made by Erects but may have been those of Heidelbergs or Mapas) form dense aggregations in the valley bottom with smaller outlying deposits, usually near springs, which to this day maintain a shallow flow down the valley. Forty-five occurrences have been recorded and the array of tools includes hand axes, cleavers, pebble tools, picks, polyhedrous hammerstones and stone knives. This very extensive series of sites has been interpreted as evidence for a wet-season dispersal of small units to higher ground with a concentration of larger numbers at a central point in the valley bottom during the dry season. This pattern corresponds closely with the seasonal movements of contemporary hunters and with the general seasonability and availability of wild resources.

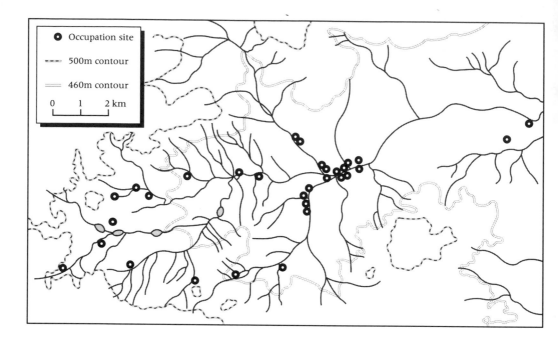

A way of life that endured for one and a half million years, spread over three continents in various habitats, cannot be portrayed by mere vignettes. Over this huge stretch of time there were few signs of significant cultural change, implying that a certain ecological equilibrium had been achieved. Nonetheless, the extensive use of fire and systematic exploitation of megafauna would have had its impact on ecosystems. What that impact might have been is discussed in the next chapter.

How long Erects lasted as a distinct type of human varied from place to place. On present evidence they survived longest in Southeast Asia, where they must have developed a distinctly different economy. If the last *erectus*-like fossil, Solo man in Java, is correctly dated to about 100,000 BP, he would have survived well into the era of anatomically Modern humans. In Africa but also especially in Eurasia, the long head start of Erect humans might have allowed them to maintain themselves in difficult or outlying localities for many tens of thousands of years; while some of these regional types could, perhaps, have interbred with later, more Modern human types, their progeny would have eventually died out and is unlikely to have influenced the genes of living people.

Replacement of one species by another is seldom instantaneous and there may be several sequences of events when potentially competitive populations come into contact. Extrapolating from other animals, a common pattern is for the more successful populations to expand over

Map of Hunsgi Valley, India, showing Acheulian occupation sites in relation to the river, springs and pools.

Skull of Erect human (dotted line) compared with early Modern (Ngaloba, Laetoli).

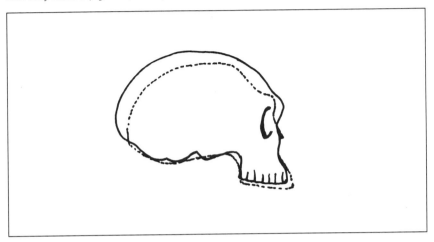

a broad front but to leave older, more established populations of the predecessor in enclaves. These tend to be areas where long-term tenancy and adaptation to local conditions give the old populations advantages. Such enclaves could be cold upland localities with peculiar food supplies or difficult terrain, where the inheritance of a complex network of known resources and refuges is not easily learnt or invaded by outsiders. Alternatively, some pockets may survive because the locality is protected by some natural barrier such as a desert or mountains.

Many overlaps between different species or types of hominid are known, but all were to a greater or lesser degree temporary. The longest coexistence was that of the big-headed herbivorous australopithecine *Paranthropus* and the smaller omnivorous *Homo*. Later, Erect humans would live in China at the same time as Mapa people (but with how much contact is unknown). It can be surmised that the more overlap there was in the use of resources (and ways of exploiting them could always be learnt faster by the new arrival), the quicker would the older population be displaced.

The best-documented displacement is that of the Neanderthals by anatomically Modern humans. Even so, the exact mechanisms are poorly understood.

The Neanderthals began as a northern population of Heidelberg in Europe and West Asia, and their elaboration of Mousterian-type tools and ornaments, burials and care for the disabled and aged, marked them

63

Neanderthal sites and their extrapolated total range. Glaciated areas and lower coast lines at about 135,000 BP are shown.

out as well-organised within a small family circle. Neanderthals are well known from excavations of numerous cave sites right across Europe. They survived the winter nights by lighting fires in caves which were close to concentrations of animals. A meat harvest sufficient to see a few individuals through the winter was best ensured by small family units living in, and presumably defending, their own caves or shelters. These were sited on animal routes or gathering grounds. A very few individuals killing, butchering and moving large carcasses had to expend more energy and possess greater strength and endurance than members of a larger group. The Neanderthals saw through winter shortages by being few and far between, living within their means and taking a mixed bag of the available animals: cave bears, mammoths, rhinos, horses, deer, reindeer, bison and ibex. They relied upon being strong, resilient and cooperative within the family, but there is no evidence that they ever gathered in any numbers.

Many populations of these massively built people would have been close to starvation during shifts in climate or during exceptionally severe winters. So long as they were the only hominids about, such vicissitudes could be absorbed but with the arrival of smaller and more numerous Modern humans, their retreat began.

Displacement of the Neanderthal by Modern man, in spite of the former's greater strength, large brain and similar range of stone tools,

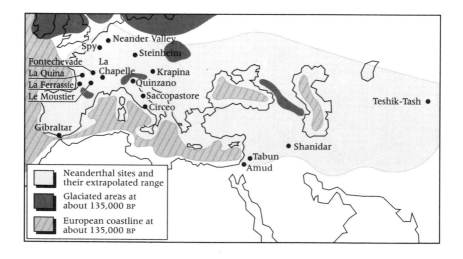

may have had many causes, including simple differences in reproductive rates, but the latter's triumph must also have been assisted by greater social flexibility and organisation.

It is unknown how closely related types of humans behaved to one another and whether any gene exchange was possible. Neanderthals and Modern humans are known to have alternated in their occupation of particular caves in some localities. In other places, it has been suggested that the Neanderthals occupied the higher, cooler hills, Moderns the lowlands.

There is now abundant evidence that the two types of humans coexisted in the same parts of southern France and the eastern Mediterranean for many thousands of years. However, as the habitat became degraded and both types of humans had to spend more time finding food, they probably came into sharper competition. An increasingly imperious pursuit of sheer survival might have hit Neanderthals harder than Modern humans if their social and cultural relationships were less flexible. In any event, by 30,000 BP the long retreat was over, Neanderthals were extinct and Modern humans would eventually colonise all the northern lands as the glaciers retreated.

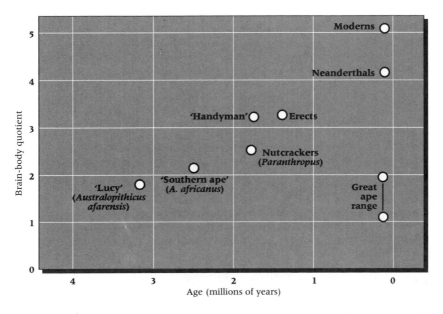

Skulls of Modern and Neanderthal compared. (left) Profile view with Modern in solid line. (right) Modern with Neanderthal half skull superimposed.

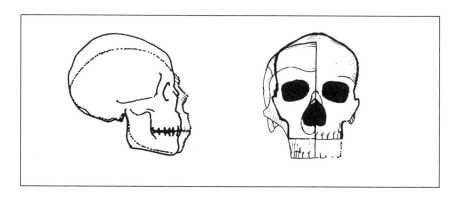

The timing of major events in the evolution of man can be summarised in a table:

TIME	EVENT
0.15 MY (?)	Modern humans expand out of Africa.
0.6 to 0.2 MY	From an unidentified source Heidelberg man spreads widely over Africa and Eurasia and gives rise to at least three regional populations (Neanderthals in Europe, Mapas in East Asia and Moderns in Africa).
1.6 to 0.5 MY	*Homo erectus* evolves in Africa and spreads to Eurasia where it forms very distinct regional populations.
2.4 to 1.5 MY	Omnivorous hominids, recognised as a new African genus, *Homo*, diversify into a complex of different forms (commonly called *Homo habilis*).
4 to 2 MY	Australopithecines diverge into predominantly omnivorous and herbivorous branches.
8 to 5 MY	*Australopithecus* emerges as the savanna form of a common and widespread omnivorous ape.

2: Eden and After

The myth of expulsion from Eden has had many interpretations; I prefer the symbolism of Everyman and Everywoman leaving the timeless garden of childhood for the realities of a finite and time-bound adulthood. If that is its imperious message as a morality play, it should be added that Eden was meant to be more than a stage of development or a state of mind. It is in the nature of myth to combine a story of naked first beginnings with the nakedness and innocence of real birth and childhood. If the beginning is childhood and Eden an atavistic playground, then Eden is home, food, comfort and security. By extension Eden is an oasis in the desert. It is even nostalgia for summer in winter. For the sick, Eden is health. Zoologists have devised an experiment where an animal is given the choice of a series of refuges, each a box where the temperature and humidity are maintained at a different level. The subject soon learns where it is most comfortable. Its box is called Eden.

The seers of an early Middle Eastern culture created a single compulsive story that melted together two elements, two intuitions about origins. One incorporates Everychild's very limited points of reference and choice of symbols for thoughts on beginnings. Here, the child's vulnerable psyche, like a rat, seeks out the most comfortable box. The other element was the universal awareness that people began in some place and some condition very different from Old Testament Arabia.

Without labouring the symbolism of Genesis beyond its inherent mythic boundaries, the concept of a Place of Beginning and later less easy peregrinations by its expellees does invite an update, drawing upon modern ideas about ecology and biogeography.

The subject can be approached by first considering the box in which hominids first felt comfortable. Climates have been central to shaping prehistory and they have been instrumental in building and breaking bridges between continents over which humans travelled to colonise lands far from Eden. Barred by their own climatic preferences, by oceans, mountains and deserts, early humans conformed to some biogeographic

*Lowered sea levels correlate with cooler tempera-
tures and greater aridity. The lowest sea levels
correspond to peak Ice Age levels, shown here
determined at Huon peninsula (but with global
application). From Chappel and Shackleton (1986),
Shackleton (1987) and Wasson and Clarke (1987).*

patterns that were not very different from those of other mammals. A brief portrait of Old World biogeography, therefore, helps to chart the dispersal of Modern humans, and some of my conclusions about that dispersal will perhaps seem less surprising in the light of other mammals and their distributions.

While the distribution of animal populations can be a guide to patterns of prehistoric human existence both are wholly subordinate to the vagaries of climate. Research has intensified in every field of human biology, with events and changes in climate in particular beginning to show some interesting correlations with human prehistory. These can now be timed with much greater accuracy than in the past.

The Ice Ages are well enough known. What is less appreciated is that they are merely a part of twenty-seven complete major climatic cycles that have taken place over the last 3.5 million years. Many of the most extreme fluctuations were during the earlier part of this period, at which time ocean temperatures were lowered, causing Africa and most other tropical areas to become much drier and the temperate zones to migrate south (so that most of the Sahara became 'Mediterranean'). One consequence of the extreme cold was to tie up huge bodies of water as ice and lower the sea levels by as much as 130 m to 150 m. This lowering of the sea created vast littoral plains, mostly fertile and well watered, which would have been important centres of human settlement. The main land extensions in the Old World were around the Mediterranean and the North Sea, Southeast Asia and the China Sea. Northern Australia and the Bering link between Asia and America were other areas where

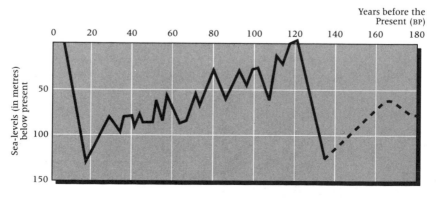

the distances between landmasses were greatly reduced as formerly flooded continental shelves were exposed. In this respect, Africa was the least affected of continents because it has such narrow shelves. Nonetheless, links with Arabia and Eurasia were broader and easier to traverse during cool periods.

A prolonged and very severe glacial period lasting about 150,000 to 132,000 BP preceded a spell of warm wet conditions in the tropics, which then cooled again about 108,000 BP and yet again about 88,000 BP. The archaeology of this period is very poorly known, yet the exodus of Moderns out of Africa and the colonisation of southern Asia would have been within this time span. The return of cold conditions and low sea levels by 64,000 BP would have reduced the area of tropical rainforests thereby improving Southeast Asia as a habitat for humans. Periods of lowered sea levels would have eased the colonisation of outlying islands and of Australia, which had been reached by about this time.

After an unstable period with some short warm spells, it got cold again about 33,000 BP. This was a period of substantial cultural change and development in several parts of the Old World, and 43,000 BP marked an advance by anatomically Modern humans into Europe where the Neanderthals became extinct by 32,000 BP. At about this time there may have been much movement and change within mainland Asia. There were certainly expansions out into the southwestern Pacific. The last and most severe phase of the Ice Ages returned about 19,000 BP and the climate has been warming ever since, except for some short blips, the last of which was the mini Ice Age of AD 1400–1850.

It is still uncertain when America was first reached. It could be as early as 30,000 BP, but the bulk of evidence points to about 15,000 BP.

All animals with extensive distributions are well adapted to a wide range of climates and humans have been no exception. For earlier forms such as Erects a large measure of physiological adaptation, plain physical toughness, must have accompanied fires and furs as their defence against cold. For later humans technology has overtaken biological adaptation.

Before the artificial use of fire by humans the ecological character of eastern Africa, where hominids first evolved, would have been very different. With fewer fires a high proportion of the vegetation would have passed through the guts of herbivores large and small, from elephants and giraffes to termites and locusts. Hominids would have had to fit in as a very subsidiary element in an ecosystem that carried many more herbivores belonging to a greater range of species than today.

East Africa is one of the most complex and diverse landscapes in the world, trenched and moulded by vast domes and tiltings, mountains, rift valleys, rivers and lakes. Forests would have threaded out from every highland and volcano, following watercourses. In the rain shadows small seasonal streams would have been lined by trees or thicket. Where the land was flat there were marshes and flood plains, mostly water-logged in the rains, but the source of much good grazing. Their margins, like those of lakes and rivers, would have been silt borders where nutrients accumulated. Here a vigorous mix of trees, thicket, termite mounds, hardpans and grassland would have had special attractions for a wide range of animals, including hominids.

A broken mosaic of trees and grassland sometimes marks the narrow transition belt between closed woody growth and open grassland. This interzone, winding between two interacting plant communities, is a narrow belt of savanna. Broaden that unstable zone to take in the ebb and flow of entire plant communities responding to inconsistent climates, and the greater part of eastern Africa would have been one huge interzone, a sprawling, wandering mosaic of trees and grasses.

It was in their accommodation to this climatic instability and ecological diversity that the earliest hominids found a niche. Before human activity became the dominant influence, the savanna mosaic would have been equally unstable, partly because of fluctuating climates, but also because the large mammal population would have fluctuated too. Their varied and vast appetites, their comings and goings would have kept the entire ecosystem in a constant state of flux; sometimes the changes were slow, sometimes dramatically quick. More consumers would have meant more plant material consumed. Contemporary animal biomass weights of up to 12,000 to 18,000 kg per sq. km have been recorded in African national parks. Such production should have been matched during peak periods in the best savanna communities of 100,000 years ago.

A majority of the contemporary species can be recognised as fossils 2–3 millions years ago and, with the ecology of living species as a guide, it can be surmised that the very varied ecological strategies of the animals rendered some of them much more vulnerable than others.

We are beginning to appreciate the impact of apparently inconsequential acts such as the daily use or lighting of fires or a preference for hunting the young of a slow-breeding animal. These quite casual activities can have very far-reaching consequences, especially when maintained over decades, centuries or more. For example, resident species tied to well-watered localities would have been more easily depleted by sustained hunting than migratory types. The former tend to be solitary, well spaced out in territories on fertile land. The latter form moving herds where sheer numbers or deliberate defensive behaviour limit the damage predators can inflict. Many forms of specialisation, especially in diet, would have tended to limit the largest animals to narrow feeding zones where they would have been vulnerable to selective hunting by humans. In this way, the spectrum of big animals would have been altered over time. The ultimate weapon, however, was human numbers. Once people were sufficiently numerous to halt or reverse natural recruitment into the prey population it became certain that species would become extinct.

The impact of prehistoric humans on African savanna ecosystems can be broadly divided into two periods. The first, effectively coinciding with Erect and Heidelberg humans, would have been characterised by a limited number of hunting techniques, probably some use of fire and a relatively low population of human hunters. Only very large, very vulnerable and specialised mammals were likely to have died out during this phase. Anatomically modern humans would have increased the pressure on animals by living at higher densities with many more subsistence and hunting techniques, more intensive firing of vegetation and by slowly expanding the range of habitats that they could inhabit. Progressively more frequent and widespread use of fire would have been the main human influence on vegetation. Today, the dominance of fire-tolerant trees, such as acacia, leadwood (*Combretum*) and wild laurel (*Terminalia*), and many fire-assisted grasses reflect hundreds of thousands of years of firing, predominantly set by humans.

The prolonged interaction between humans and African savanna communities must have established cultural and technical traditions that were so closely tied in with behaviour and psychology that they were not quickly abandoned even in non-African habitats. African savanna ecosystems evolved in tandem with *Homo*. On the one hand, the animals and plants evolved some defences against human depredation.

On the other, a complex of endemic diseases, competition, predators and social and environmental constraints probably combined to keep human populations in some sort of balance. The African Eden was no paradise, but it was where hominids had been part of the landscape for at least 5 or 6 million years.

Emigrant hominids would have had a preference for habitats with similar climates, fauna and flora where these traditions could be maintained. Parts of the Middle East and India have the closest ecological similarities to eastern Africa and, for each successive wave of humans coming out of Africa, this great arc of country to the northeast would have been the first to be colonised and populated. It is frustrating that human fossils are almost non-existent here in spite of numerous artefact sites which cover a wide span of time.

The likely timing and extent of hominid dispersals are summarised in tabular form on p. 66. Barriers to the dispersal of a terrestrial animal are seas, mountains, deserts and other inhospitable habitats. Since climates change and sea levels fluctuate, populations can find that the gate opens for a few hundred or thousand years and then closes again. The consequence of any interruption of gene flow is the formation of separate gene pools, and it is in the context of early human gene pools that Old World biogeography has a special interest.

Within the continent of Africa, divisions are ecological. The broad belt of equatorial forest has seldom, if ever, bisected the savanna from the Atlantic to the Indian Ocean, because drier corridors have always run from north to south. Nonetheless, there are numerous differences between the fauna and flora of northern and southern savannas. The southern and eastern savannas are much more varied and complex and, therefore, support a much wider variety of species, many more antelopes, zebras, rodents and small carnivores. The northern savannas, simpler and sandwiched between oscillating desert and forest boundaries, have a poorer fauna, but there are some distinctive species, such as the patas monkey, a type of ground squirrel and several species or races of antelope (each with a close relative to the south).

This evidence for separate gene pools in the southern and northern savannas may be relevant for early prehistoric humans. Such a division among humans need not only have depended upon their following a widespread mammal pattern, but diet, population density and subsist-

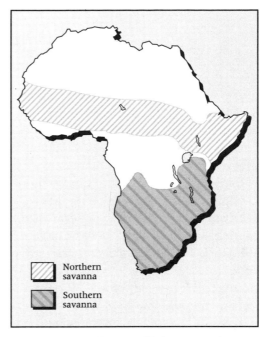

Distinct northern and southern savannas. Margins of intervening forest would have fluctuated with climatic change.

Northern savanna

Southern savanna

ence technologies could also have differed substantially between the two areas. In humans the genetic distinctness of regional populations would tend to be reinforced by such cultural differences.

When Africa is examined specifically as the habitat of more developed humans, it can still be analysed in terms of discrete units, because there are physical or ecological discontinuities and several 'narrows' or weirs which would have broken up populations and encouraged the emergence of differences between them. Every phase of innovation in human evolution is likely to have this biogeographic dimension, because these natural subdivisions within the continent provided different and varied theatres for the development of special new skills and faculties.

To extract patterns that are relevant to humans it is necessary to identify both the factors that favour and those that discourage human occupation. The field can be narrowed immediately by excluding dense, closed forests and true waterless deserts. Less obvious are areas that are too infertile, water-starved and monotonous to have significant resources for people. High, cold uplands were also unattractive. There are many such areas – they are capable of supporting scattered pockets of people (just as the central Kalahari Desert does today) but they are not, especially during dry periods, prime human habitat. The positive indications needed to identify the latter are an equable climate, adequate water, fertile soils, a numerous and diverse fauna and flora, a mosaic of different vegetation types – a varied landscape. When Africa is mapped in this way there are four main blocks, running from the mouth of the Red Sea to the Cape of Good Hope. The core area is East Africa, comprising today's Tanzania and central Kenya. In terms of mammals, especially large ones, this is the richest region on earth in numbers of species, diversity and abundance. This richness relates to East Africa's volcanic fertility, climate, geographic diversity and position and its evolutionary history.

Fossils from East African deposits confirm that this diversity has long been characteristic of the area and has made it particularly favourable for a very wide range of predators, including humans. North of this richly endowed land lies the Ethiopian zone. The main human habitat here (until relatively recently) was the narrow but very long Rift Valley floor. This has much in common with parts of the East African Highlands but has always had a much less varied fauna and flora. East Africa's southern edge is bottlenecked into an escarpment and cool uplands between the southern tip of Lake Tanganyika and northern Lake Malawi. East of Lake Malawi, Mozambique forms a sort of southern annex to East Africa.

I call the area immediately southwest of this the Broken Hill zone (to commemorate a fine hominid skull). This covers the whole of today's Zambia, Zimbabwe, Malawi and some bits of other countries as well. The Broken Hill zone has less variety, both ecologically and physically, than East Africa yet carries a rich and diverse fauna within a mosaic of streams, rivers, floodplains, rocky outcroppings, forest, woodlands and swamps. Drier, more impoverished land, including the Kalahari Desert, lies to the west.

The broad Limpopo Valley marks the beginning of the entirely different South African zone. This resembles East Africa in having an exceptionally lively topography and a very diverse fauna and flora. Nonetheless, its cool uplands tend to be infertile in places and seasonal extremes force many of the herbivores to be migratory. Its rich and varied fauna is more patchy in distribution and its climate, especially inland, is more fickle and extreme than in East Africa. Contrasting climates and a more difficult subsistence in the south have probably influenced differences between contemporaneous hominid fossils from South and East Africa, whether they are *Australopithecus* or *Homo*. In almost every instance where the two can be properly compared there are subspecific or species differences.

Early hominid centres of distribution.

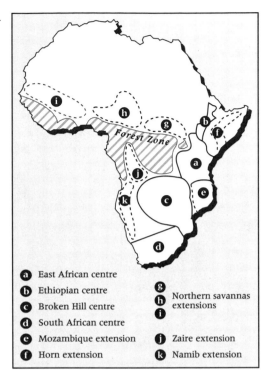

- **a** East African centre
- **b** Ethiopian centre
- **c** Broken Hill centre
- **d** South African centre
- **e** Mozambique extension
- **f** Horn extension
- **g** **h** **i** Northern savannas extensions
- **j** Zaire extension
- **k** Namib extension

There are subsidiary outliers to the four zones just described, but in every case these are areas that would have been subject to much greater ecological instability. Nonetheless, these areas can be identified on similar criteria to the eastern ones and the positions and extent of seven such outliers are shown above.

Specific adaptive changes or modifications to behaviour and technology may well have developed in response to local conditions in each of the four zones but very important distinctions are involved. Wherever a development was related to local conditions, such as altitude in Ethiopia, cold in South Africa or flood-plain living in Broken Hill, that capability, whether genetic or not, was unlikely to spread widely outside its centre of origin. It was this sort of differentiation that tended to make for distinct southern and eastern races of hominids.

By contrast, any adaptation that conferred a universal and decisive advantage would eventually spread because, in spite of the narrow gateways between zones, there is a broader ecological continuity throughout the uplands of eastern and southern Africa. The immediate consequences of more efficient and reliable hunting, of increasing the harvest, would have been better-fed hunters with more surviving children and an expanding population.

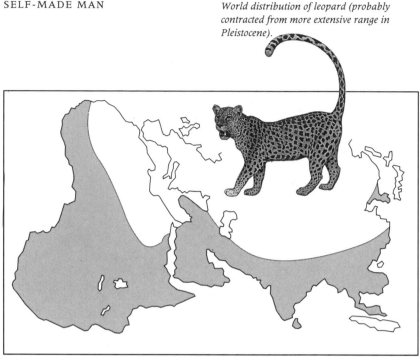

World distribution of leopard (probably contracted from more extensive range in Pleistocene).

In the later stages of human evolution such genetic advances were closely linked with technological innovations and they eventually spread far beyond Africa.

Of course, barriers within the continent are dwarfed by the desert and sea barriers between Eurasia and Africa. Even so, many terrestrial animals have crossed them at various times in both directions. Consider some patterns. Living at the top of the food chain, like Modern humans, some carnivores have escaped confinement to a single continent. The leopard, for instance, is so versatile in its choice of animal food that it can live in near-deserts, moist forests and most habitats in between. It was once found throughout Africa and southern and eastern Asia. Its dietary versatility allowed it, like humans, to be relatively cosmopolitan.

Lions and cheetahs also occurred outside Africa, but both had a more limited choice of prey animals and more specialised hunting techniques and so were less widespread. Striped hyaenas, honey badgers and caracal lynxes are other African carnivores that still range widely in the drier parts of Asia for similar reasons; until recently the spotted hyaena was even more widespread.

Going back millions of years, Africa was the original source of the ape

76

and elephant families, both of which later flourished in Eurasia. Flowing into Africa from Eurasia were various rodents, early antelopes and giraffes, rhinos and horses, to name but a few. Today, zebras are regarded as typically African, yet they originally evolved in Texas and reached Africa only after crossing the whole of Asia. A relative of the common zebra survived in India until mid-Pleistocene times, and zebras once ranged across large parts of what is now the Sahara.

The distributions of such animals are relevant to human evolution, because they show that there are three major routes of diffusion between

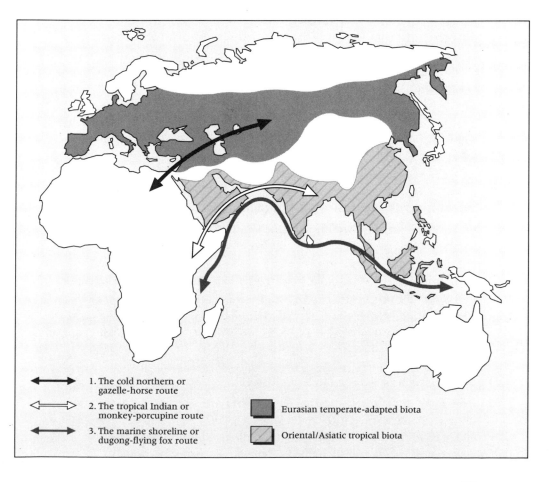

1. The cold northern or gazelle-horse route
2. The tropical Indian or monkey-porcupine route
3. The marine shoreline or dugong-flying fox route

Eurasian temperate-adapted biota

Oriental/Asiatic tropical biota

Africa and Asia, each with a totally different ecology. First there is the northern steppe, what could be called the gazelle-horse connection. Second there is the tropical Indian or monkey-porcupine connection. The Himalayas and its outliers are of course the dividing line between these two tracts. The third, more sinuous link is the marine shoreline or dugong–flying fox connection.

The first route was followed by zebras and asses diffusing across Asia *to* Africa and by gazelles, cheetahs and others eastwards *out* of Africa. The many mountain ranges and desert plains along this route tend to break its continuity, especially during periods of cold or drought.

Although the drier areas of central Asia have some ecological resemblance with arid Africa, all its endemic fauna and flora show signs of adaptation to cold drought, not the hot drought of Africa, and its most typical grazers are well-fleeced sheep and goats rather than thin-coated tropical antelopes.

The tropical mainland between Africa and the Far East is even more interrupted by large mountains, rivers and plains. Although there are distinct fauna and flora with rather few species in common, there are numerous closely related groups with representatives in Africa and Asia. Thus, the monkeys, mongooses, porcupines, squirrels, pangolins, peacocks and many others have diffused both ways (some of them many millions of years ago) and have speciated behind the numerous barriers that intervene. Nonetheless, there are the wide-ranging leopards, jackals and honey-badgers to remind us that some tenuous degree of gene-flow can be maintained overland in southern Asia.

Further east, the humid-adapted flora and fauna are locked into the equatorial Southeast Asian archipelago and islands. Orang-utans, tapirs and flying lemurs typify the Oriental forest community. No terrestrial, forest-adapted species has succeeded in moving west by overland routes during the period of human evolution, because there have always been arid barriers to cross.

The two land routes across Asia are particularly relevant to the dispersal of early humans in Asia. Although there are virtually no fossils from intermediate localities, it is possible that the two known populations of Erects had diverged into northern Asiatic and tropical Southeast Asian populations as a part of the process of diffusing out of Africa.

A steppe route could have followed the drainage lines that flow northwards out of the great mountain chains between Persia and the Tibetan plateau. However, this is conceivable only during a warm, wet period. The southern route might have embraced much of the Indian subcontinent.

A dividing of populations by the Himalayas could, therefore, have resulted in the prime habitats of the Far East being colonised from two directions. If so, such colonists would have reached the Pacific having developed totally different cultural traditions, because one had become adapted to open steppes and the other to tropical wooded environments.

Anatomically modern humans, as essentially tropical animals, would have followed the Indian route. A great unknown is how Moderns, Mapas (best visualised as Oriental Heidelbergs) and Erects might have interacted within these vast territories. The array of fossils from China suggests that two, perhaps three, types of humans were about in the region during the Middle and Upper Pleistocene.

The third, seashore route is a narrow realm where terrestrial and marine ecosystems meet and sometimes overlap. Furthermore, it is a pathway where most of the traffic flows from the east. Recent overland invasions of Africa from the east have been inhibited by deserts and seas, but the seashore thickets have helped several east Asian birds reach Africa together with various coastal plants. Thus, Indian robins, a sunbird, and a leaf-nest-sewing tailor bird hedge-hopped westward this way.

Islands off the east African coast are few and far between but they have been colonised by Indonesian bats, flying foxes, which live in vast and spectacular colonies and roost by preference close to the seashore. This preference is so marked that it is possible that the bats are physiologically dependent on drinking sea water for certain trace elements.

The coasts of the Indian Ocean and South Pacific are ecologically very similar from the Solomon Islands to the Mozambique Channel, but it is only in the Far East that the shoreline fragments into thousands of islands in shallow seas. The inshore shallows of these tropical oceans once supported a large grazing mammal, the dugong or sea cow, which existed in vast numbers, especially in the seas of Indonesia and the South Pacific. The dugong was the largest member of a very abundant and recognisable community of shoreline herbivores that included about five species of turtles, fish, crustaceans and many other organisms sustained or sheltered by extensive beds of seaweed, especially sea grasses growing in sunlit shallows.

Although the dugong ranges very widely, it suffers constraints. It exists within very narrow temperature limits and depends on an abundance of grazing that is sensitive to disturbance. It is unable to tolerate consistently rough seas. Furthermore, its linear habitat is intrinsically different from a two-dimensional one. Contacts between small, strung-out populations are tenuous and any variations in climate or resources will make survival more precarious. Thus populations build up only where the coastline webs (as around islands) or expands, as in broad, shallow bays or straits. The pattern of decline in dugong populations suggests that their long-term survival has depended upon just such 'expanded' coastal localities where large nuclear populations can be maintained.

Cultures that were specialised for shore-dwelling could have followed their prey, the dugong, in a preference for island archipelagos and bays with an extensive framework of coasts, shorelines and shallows. Wherever the hinterland permitted it and wherever the incentives were strong enough, seashore dwelling communities could have been beckoned back inland. Movement in the opposite direction, from inland to shore, would have been common enough too but once seashore dwelling had become a specialised cultural niche, generalised mainlanders would have had fewer opportunities to intrude.

Prehistoric middens (and the historic traditions of contemporary peoples) show that many small communities used to make seasonal movements between inland or riverine camps and coastal sites. Once again there are animal precedents for this. Hyaenas regularly patrol the bleak desert foreshore in southwest Africa and lions arrive during the fur-seal breeding season to grow fat upon the thousands of pups in the rookery. Foxes, jackals and cats find seasonal bonanzas in sea-bird colonies.

There is a larger evolutionary dimension to this, because it was the rich resources of the seashore that first attracted creodonts there; they ended up as whales and dolphins. Later carnivores became sea lions; yet other seals, and still more recently the weasel family adapted to become otters.

In every case, it was the immediately accessible food supply that must have induced shore dwelling in the first place but, as the waters beckoned, the animals went in deeper. Such a radical change of habitat involved comprehensive changes in anatomy and physiology, so that the animals were effectively transformed. Even so, each lineage changed according to its own unique pattern of adaptation. The elephant lineages that became sea cows, the reptiles that became crocodiles, the innumerable rodents, insectivores, ungulates, mongooses and others that have become secondarily aquatic or semiaquatic have all been 'transformed' in some unique and special way to become beavers, desmans or hippopotamuses.

No primate has crossed that boundary but several species exploit the water's edge. Crab-eating macaques live up to their name in some localities. Vervet monkeys flourish in some African mangrove swamps, while proboscis monkeys in Borneo live nowhere else and regularly wade and swim between banks and islands.

A theory that made a brief flourish in the 1960s held that the nakedness, subcutaneous fat and peculiar distribution of hair on humans could be explained by a brief flirtation on the part of our ancestors with 'going aquatic'. The theory had its beginnings with a lecture to a subaqua club by Alistair Hardy, professor of Zoology at Oxford University. The idea was taken up by others, and the barren Red Sea coastal islands were postulated as prime candidates for the site where 'aquatic apes' developed! This way of life was then conveniently abandoned under the influence of climatic change.

This eccentric theory had the virtue of drawing attention to lakes and seashores as prime habitats not only for our immediate ancestors, but also for still earlier members of our lineage, scavenging, versatile apes and apemen. Although australopithecines undoubtedly reached the sea beside some major estuaries, their main habitat was beside rivers, lakes and water holes.

All hominids would have depended on fresh water to drink, so that the predominantly dry coastline of Africa would have been very literally a marginal habitat for humans and prehumans. This marginality makes the Red Sea an impossible habitat for the aquatic hypothesis. In Asia, similar considerations would have applied. Well-watered woodlands, plains or savannas throughout southern Asia would have resembled those of Africa, and a succession of human populations followed one another in occupying the vastness of Persia, the Indian peninsula and lowland China. For these continental masses the coasts were mainly a fringe.

In the more tropical southeast the situation was rather different. Dense tropical jungle is a very difficult place to live in without advanced tools. Although Erect humans invaded the Indonesian peninsula, reaching as far as Java (which during periods of lower sea levels was joined to the mainland), they probably concentrated along river valleys in the drier, more open areas. Notwithstanding the instability of the coastlines, this part of the world is where the interface between sea and land is at its most fragmented. Nowhere else in the world are there so many islands, nowhere else are there so many thousands of kilometres of coastline.

The same heavy rainfall that rendered so much of this region uninhabitable jungle also ended up flowing down innumerable rivers and streams, thus multiplying opportunities for waterside animals to find a living. Humans who lived in Indonesia for over a million years had an environment that was strikingly different from that of their northern counterparts. Not surprisingly Javan *erectus* diverged into a different race (some would say species) from the 'Peking man' of China.

Although Erects are not a central concern here, their earlier development of a tropical waterside existence (and their physical presence when *Homo sapiens* first arrived) would have presented the newcomer with some ready-made traditions in local survivorship and, of course, with competitors. For *Homo sapiens* in southeast Asia the best habitats were in the rare drier and more open localities, beside rivers and the sea. The latter habitat was potentially the richest in food resources.

A second by-product of the 'aquatic ape' theory was that it emphasised that physiological and other changes are very necessary if any organism is to move into such a different sphere of existence. Unlike us, they had no wet suits, goggles and other protective devices. Yet there is abundant evidence that the temptations that took the ancestors of whales, seals, hippos, otters and beavers across this boundary were equally strong for some populations of human beings. The boundary-crossing that totally transformed the ecology and the very nature of the New World, Australia and eventually all the Oceanic islands was achieved by humans paddling flimsy rafts or canoes across the gaps between continents.

Consider some of the implications of boundary-crossing or invasion for the invader and the invaded. Every continent has a fauna and flora that is the sum total of many invasions and later proliferations, but the new arrivals (particularly successful colonists) were seldom simple additions. It is well known that dominant intruders set off lasting changes and prolonged repercussions in the ecological communities that they invade. For example, fossils have revealed that many archaic endemic mammals of South America quickly succumbed when a new land bridge brought modern mammals in from North America some two million years ago.

Extinction is also a modern process. The last of the giant St Helena earwigs are thought to have been eaten by alien house mice in 1988, but their demise was a mere detail in the protracted demolition of an entire and unique island community by immigrants possessed of what might be called 'continental vigour'. Most people know that sheep, cattle, foxes and rabbits, all 'vigorous continentals', have changed the landscape and ecology of Australia. Fewer are aware of the scale of change that attended the much earlier arrival of humans in that island continent.

Resilience and adaptability are exact and specific properties of an organism. They have evolved in response to the challenges that have been met during the evolutionary life of an entire lineage. No organism can be expected to have genetic defences against a challenge its ancestors never met. When a new challenge is faced there are three possible outcomes. The first is that existing defences are sufficiently plastic to provide protection. Second, natural selection favours minor modifications to existing structures or defences so that surviving populations become better protected than earlier ones. Third, there are no defences and the population becomes extinct. This may take a generation or two, the attrition may be spread out over centuries or the collapse coincides quite simply with offtake by humans exceeding recruitment in the prey animal.

The speed with which very large defenceless animals can be exterminated can be illustrated with an historic example. Steller's sea cow (*Hydrodamalis*) originally ranged throughout the shallower inshore waters of the North Pacific, where its ten-ton bulk was sustained by feeding on seaweed. Over all their accessible inshore range, sea cows were probably eliminated by the earliest prehistoric sailors along this coast. They survived, however, on two remote islands more than 200 km

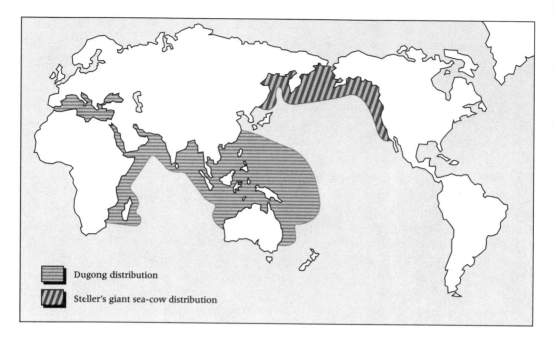

Dugong distribution

Steller's giant sea-cow distribution

off Kamchatka, where they were discovered by Russian sailors in 1741. Their defensive behaviour resembled that of the dugong, in that males would interpose themselves or try to buffet fishermen catching a female. The defensive behaviour of these 7.5 m sea cows merely accelerated their extinction. This was how Steller described it:

> If one of them was harpooned all the others tried to save him. Some formed a ring round their wounded comrade and endeavoured in this manner to keep him from the shore. We also observed with astonishment that a male came on two successive days to his dead mate lying on the beach as though to enquire after her wellbeing.

In spite of the herds being large and spread along some 300 km of shore, the last animal was killed in 1768, twenty-seven years after the species was first discovered. There are now no large-scale consumers of seaweed and kelp. The ecological consequences of the sea cow extinction must have been immense but are completely unknown.

The extinction of Steller's sea cow is unusual only in being sufficiently recent for the dates to have been precisely recorded. The celebrated dodos of Mauritius and Rodriguez are another case of island vertebrates becoming rapidly extinct through a combination of direct human predation supplemented by that of their pigs, dogs and rats. It is well documented that this quartet, with cats and goats, have done more to

84

(left) Map of dugong distribution in prehistoric times and of Steller's giant sea cow in the North Pacific

(right) Moa and Kiwi.

exterminate small oceanic islands' fauna and flora than all manner of natural predators, catastrophes and climatic changes. In the dodos, various endemic plants of Mauritius would have lost a major disperser and seed processor. At least one plant is thought to have declined as a consequence, but the ecology of Mauritius is too shattered for any reliable reconstruction to be made.

While dodos were far and away the largest birds in the Mascarene Islands, they were too recently evolved to have totally disguised their origins as the descendants of flying doves. In New Zealand, the traits that typify dodos – loss of flight, enlarged size and extreme abundance – were taken much further by some twelve species and two families of giant birds, the moas, which weighed between 10 and 200 kg and still puzzle ornithologists as to their origins among flighted birds.

New Zealand has been isolated from other landmasses for millions of years and has no terrestrial animals, except for those that could fly or swim there, or small animals of such primitive origins that they rafted on the landmass itself. The decline of the moas began with the arrival of humans, dogs and rats about AD 1000. With so many moa species it is uncertain how long it took for them all to become extinct; the larger ones were gone within 300 years, the smaller after about 500 years. Scientists in New Zealand (G. Caughley and A. Anderson) have worked out that the moas formed a guild of browsers and fruit-eaters that were able to sustain high densities by cropping the vegetation in a very heavy and systematic way. Moa bones were sufficiently dense in places to make ploughing difficult for the early European settlers, who sent them by the wagonload to specially set-up bone mills.

So abundant were the moas to begin with that Maori hunters took the meat-bearing legs and threw necks, heads and other offal away. The

Maori brought tropical, Polynesian traditions to this temperate island and a strongly marine economy, so the first summer camps were coastal and exploited populations of fur-seals, dolphins, fish and molluscs that had no previous experience of being hunted or collected. In one site which represented the harvests of about twelve summers, moas made up about one sixth of the total meat eaten. Another riverine moa-hunting camp further inland covered about 16 hectares and was scattered with earth ovens and the remains of several moa species.

The anthropologist W. Shawcross has made detailed comparisons between the early Moa hunting sites and later ones, where they were becoming scarcer, and found interesting changes. At first a group of families, numbering about fifty people, could range over the uninhabited landscape as a coherent foraging band. Later sites showed summer dispersals and winter gatherings, but human populations then began to fluctuate, with high-density villages at some times and a scattering of small settlements at others. This suggests that there was an initial build-up of numbers fuelled by easy harvests of seafood and moas. This phase had no need for agriculture (but their dogs, which were bred for strong necks and jaws and are known to have run down flightless birds, also represented a form of stock and were eaten). A strong coastal bias remained, but there may have been as many as 30,000 people living inland and agriculture became necessary as prey became scarcer.

It has been suggested that bracken rhizomes became a larger source of food for some late Maori settlers than their cultivated crops. If so, this represents a sorry contrast with their early affluence. However, a special interest of this is that bracken is a fire-climax plant. Without people to start them and with moas consuming a large proportion of green growth, fires were rare in New Zealand before humans arrived and the indigenous forest had no tolerance or resistance to fire. Podocarpus pines were widespread, probably because they benefited from the moas' preferential cropping of broad-leaved undergrowth and their dispersal of the pines' fleshy seeds. With the decline and end of the moas' browsing, dense broad-leaf growth shaded out young podocarps and suppressed regeneration. Fuelled by vegetation that was adapted to heavy browsing rather than fire, torching by humans resulted in an almost total deforestation of the eastern half of New Zealand. Bracken replaced moas both physically and as a central item of human diet.

This denudation began within about a century of the moa hunters' heyday, and the very last moas could have perished in bush fires as frequently as on Maori spears.

Sea cows, moas and dodos are very recent extinctions but the sequence of events in New Zealand has illustrated a process that would have operated during earlier colonisations of continents and islands. Fauna and flora that were innocent of previous contact with ex-African hominids are commonly called 'naive'. Long before the New Zealand cycle, naive communities in Australia, the Americas and Madagascar fed the first few generations of healthy, fertile human pioneers. The ecological upheavals that followed always precipitated a series of changes.

Any animal that is sufficiently numerous to contribute significantly to the diet of a burgeoning population of hunters cannot be a trivial component in the local ecosystem. The moa-to-bracken sequence serves to illustrate the paradox that creaming off the dominant herbivores is not necessarily of benefit to the plants they used to feed on. A relatively stable relationship between consumer and consumed can only evolve over a very long period. The forces that break that relationship have their own dynamics and, for the new human-induced ecosystems, that usually involved his prime tool, fire; and it takes a specialised plant to cope with being regularly burnt.

For the over-numerous descendants of those who exploited a naive fauna and flora so carelessly many cultural and economic changes became necessary once the naives were gone. Later human settlers, like the Maori, fell back on farming but the earlier colonists of America and Australia had to turn to less naive endemics and more difficult, usually smaller foods. At this stage necessity may have forced people to learn complex processes of washing, leaching, fermenting and parboiling otherwise poisonous foods.

On most islands the sharp drop in resources meant a corresponding decline in population and the agents of that decline were often cruel. For some there was migration, for others starvation or cannibalism and warfare but most likely of all was poor health, disease and many deaths. The stone Moai statues on barren Easter Island stand witness to the abundant energies of new colonists on a once forested subtropical island. Their solitude on windswept steppelike slopes speaks of the consequences.

It would be tedious to list the known casualties of this process for all the islands and continents that humans have colonised, but it is important to highlight some of the major continental differences, to 'take measure' and thus appreciate the scale of past extinctions.

The African fauna, after its early losses, was neither naive nor static. Although many species have shown great durability, their genotypes are probably very different from what they were a million years ago. They are the result of extremely intense competition in the most crowded and diverse of all mammal communities.

Southern Asia resembled Africa in losing many of its largest specialised animals at a relatively early stage. Erect humans shared Java with two species of elephant, two rhinoceroses, the peculiar tapirlike *Nestoritherium*, sabre-tooths, black bears, leopards and the (now exclusively African) spotted hyaena. There were also many minor carnivores, which are good indicators of a rich and diverse fauna of smaller prey animals.

The Southeast Asian islands had small elephantine *Stegodons*, rhinos, various pigs and monkeys and, among the carnivores, otters. It is interesting that among the Pliocene fossils of southern Asia is a giant mollusc- and crab-eating otter called *Enhydrion*, which was larger than the Pacific sea otter. Anatomically, *Enhydrion* was intermediate between the living African clawless otter, *Aonyx*, and the sea otter and it may also have occupied an intermediate ecological niche. Sea otters must have had fresh water antecedents similar to *Aonyx* and *Enhydrion*. The inducement that would have drawn a tropical riverine animal into temperate and eventually arctic seas would have been the enormous resources of estuaries and rivermouth seashores. Here, there were abundant molluscs, crabs, crays, turtles and other easily harvested but tough-shelled animals. Given that these resources are still abundant, why has this otter become extinct while its marine and African cousins survive?

A necessary prerequisite of gigantism is a large and reliable food source, and gigantism can be a form of insurance against predators and competitors. Where humans are the predator and competitor, large size becomes a liability and it is quite possible that *Enhydrion* was one more 'naive' casualty of humans when they took to exploiting river and estuary resources in a big way.

Southeast Asian fauna demonstrates very well the role of isolation in the evolution of unique forms of animals. The continental forms tend to

A sample of some Australian vertebrates that became extinct during the last 60 KY. Upper two rows, giant members of the wombat and possum groups. Middle three rows, giant kangaroos. Bottom, extinct echidnas, devil, giant emu, megapodes, giant goanna, giant python (After P. Martin and J. Klein, 1984).

be more resistant to extinction unless they are very large. The island forms are more vulnerable for many reasons, not the least being their smaller numbers. The lack of exposure to predators, which often die out quite quickly on islands, becomes a part of the total ecological 'climate' to which species adapt. Thus island endemics are frequently naive. Islands like Mindanao in the Philippines have many endemics (for example, 79 per cent of all mammals). Borneo, which had many periods of connection, has 25 per cent endemics, whereas the Malay peninsula has only 2.7 per cent.

The most spectacular extinctions followed human arrival in Australasia, where marsupials had radiated into a wide variety of giant forms, the majority of which have disappeared. Of thirty-seven extinct herbivores, seventeen had deep mill-like teeth that could chew coarse grasses and other rough vegetation. In spite of a relatively poor fossil record (about

forty relevant localities), the giant marsupials were clearly abundant and widespread; the largest, *Diprotodon optatus*, stood nearly two metres at the shoulder and would have weighed well over a ton. It was a sort of browsing, rhinoceros-sized wombat. Another, *Zygomaturus*, had a form of blunt nasal horn or nose pad (with which it probably jousted). With a body about two metres long, this too was a lumbering rhinolike beast but it may also have been able to excavate shallow roots with its large nailed forefeet. The range of types included somewhat piglike forms, giant rodentlike animals and even a sort of marsupial chalicothere or giant sloth. All these were heavy animals. Australia's bovid equivalents, the kangaroos, ranged from heavyweights nearly three metres long to small species such as the (still surviving) agile and hare wallabies.

The heavyweights may well have been resident species, living only in the best and most fertile regions, while the more mobile species were more likely to have been nomadic. Large size is nearly always linked with extensive ranges and big appetites. Many large herbivores, especially those with primitive metabolisms, would have had to spend so much of their time munching vegetation that they would have been particularly ill-equipped to cope with the attacks of a fast, mobile and well-armed predator. The marsupial grazers were supplemented by two giant tortoises and an emu that would have weighed over 100 kg.

The main predators, then as now, were crocodiles; one was a giant, inland freshwater type, the other a 200 kg terrestrial dry-country crocodile. These species were smaller than the seven-metre-long monitor or giant Komodo dragon (*Megalania*). A very thick five-metre python, *Wonambi*, was also common. Two large carnivorous marsupials were the giant possum (*Thylacoleo*), which seems to have combined big cat, hyaena and bear-like features and the *Thylacine*, or marsupial wolf.

There have been attempts in Australia to assert that climate killed the megafauna. The climate lobby argues that large animals are vulnerable to climatic change, because the bigger the animal the more slowly it reproduces, so that its genetic responses to change in the environment are too slow to allow it to adapt; it is supposedly lacking in adaptive flexibility. However, while it is true that big bodies can be a liability, they permit a wide range and hence a larger food supply and, even more important, they normally confer a high measure of protection against both predators and extremes of temperature.

The suggestion that large animals are inflexible is contradicted by the demonstrable success of large mammals in adapting very quickly to less food on islands and it is interesting that they do this by getting smaller. Elephants, hippos and giant pigs evolved pygmy forms on many islands where they flourished until humans appeared. Very large animals may well have declined in numbers during periods of climatic change. They might have retreated into relatively small enclaves, but few would have gone extinct because they had already been through several eras of drought and cold before humans arrived.

The clearest indication that humans killed these animals comes from Mammoth Cave in Western Australia, a site that has been dated to about 37,000 BP. Here there are remains of deliberately broken and burnt animal bones, including giant kangaroos, giant spiny anteaters and the giant possum. The bountiful harvest of large, 'naive' animals would have encouraged the first human arrivals to spread widely and multiply. Although there would have been cultural distinctions between seashore dwellers and inlanders, the beneficiaries of the first bonanza could have developed and diffused a relatively uniform large-animal hunting culture all over the continent. With the decline of all but the most agile marsupials and a major upheaval in the vegetation, a greater variety of locally appropriate or regional strategies for survival is likely to have developed.

The fact that there were ecological changes is most graphically demonstrated in rock art from the Kakadu escarpment, where older representations of arid-adapted kangaroos have pictures of waterfowl and barramundi perch superimposed on them. These were records of a climatic rather than human-induced change. It is most likely that the littoral people shifted with their habitat, while inland people also ebbed and flowed with the distribution of their favoured foods.

Cultural fragmentation was reflected by the great diversity of cultures and languages of the Aborigines before they were subjugated by European settlers. In spite of an absence of physical barriers Australia has well-marked ecological zones and the genetics of Aborigines confirms a high degree of differentiation within Australia, coinciding broadly with ecological regions. For so much differentiation to have taken place, a long time must have passed in a remarkable degree of culturally maintained isolation.

The one animal that remained an important quarry in most parts of Australia was the kangaroo. The living red and eastern grey kangaroos were originally one-third larger than they are today. The most plausible explanation for this adaptive shrinkage is that small animals need less to eat, tend to be faster and more agile and mature and breed faster, so that they are better able to offset the depredations of hunters. Smaller kangaroos have shown more modest shrinkage over the same period but there is one very telling exception, which argues very strongly for human predation being a selective influence on the body-size of kangaroos. The Australian zoologist and palaeontologist Tim Flannery found that the red-necked roos (*Macropus fuliginosus*) of Kangaroo Island actually showed a 7 per cent growth during recent millennia. After their original settlement, humans died out on this island and are known to have been absent between at least 4,000 to 2,500 BP. Flannery believes that the enlargement in the red-necks' size reflects their respite from intensive hunting over this period.

A higher proportion of large animals became extinct in Australia than in any other continent. On present knowledge of fossil fauna (which is certainly incomplete) 86 per cent of Australia's megafauna (all animals calculated to weigh over 44 kg) is lost. South America is not far behind with 80 per cent and North America with 73 per cent. In the latter more than seventy genera of large mammals became extinct over a very short period. The great majority, perhaps all of them, disappeared shortly after the arrival of modern humans and there are abundant kill-sites to show that the first Americans had a gigantic meat-eating bonanza.

By contrast, Africa has been estimated to have lost 15 per cent and these were mostly the more specialised relatives of types that have survived. It was the most aberrant and perhaps over-specialised animals that disappeared much earlier and, if their loss was influenced by hominids, as seems plausible, *H. erectus* was the predator. For the most part, it was relatively late humans (between 40,000 BP and the present) with their very highly refined weapons and large numbers that were mainly responsible for later losses. When the pattern of extinctions is looked at worldwide, the first great spasms followed very quickly on the first arrival of Modern humans (with a possible early rehearsal by *erectus* in Africa and Eurasia). Another great wave of extinctions accompanied the spread of Europeans with their boats, weapons and attendant rats, cats, dogs, pigs and so on. Another phase of extinction has now begun, its causes are not so different from the other ones but its scale may be the largest yet.

Contrary to the sentimental image of a life in harmony with nature, people have always taken a vast delight in the display and practice of techniques, especially novel ones or ones unique to a culture or economy. The manufacture, fine tuning, placing, maintenance and application of technology would have been the main preoccupation of prehistoric people no less than of today's boffins. The effect of all this technical ingenuity was to increase the number of predatory niches that humans could occupy, but it is only with the hindsights offered by ecology that this becomes apparent.

For the large mammals that had evolved so many distinct forms in different parts of the world, human expansion into their areas of endemism was a terminal disaster. It was followed by the less well known but equally fundamental changes in vegetation that were illustrated in the tale of moas and bracken. Now with an implacable momentum agriculture is transforming or replacing all but vestiges of the earth's original natural ecosystems.

For the first time ever, science has begun to develop the intellectual and analytical tools to study and reconstruct human and natural history as they were. The principal materials for that study are not so much stone tools and fossil bones as the vestigial patterns of once vital and autonomous species and the interrelationships of entire ecological communities. Those vestiges are currently under threat from the modern equivalent of prehistoric man enjoying a bonanza while it lasts. Their tools are money and the mega-machines it can buy, their costumes striped suits and spectacles, not ochre and skins. Curbing their appetites is going to be an essential prerequisite in rediscovering Eden.

3: Adam Abroad

The West Indian islands are exactly half a world away from India, and real Indians lived five times further away from Spain than Columbus and his queen had hoped back in 1492. Yet the inertia and conservatism of language are such that Caribbeans and Americans are still called Indians. The term 'West Indies' is not only a verbal monument to assumptions that would shrink the earth in half; the name also proves the psychological need for cloaking the unknown with the known, with something familiar. Under that cloak having a named people inhabit a named place has been knowledge enough.

Today's voyages of discovery seek to establish when, how and from where the Amerindians got to America, the Aborigines to Australia and the Europeans to Europe. The quest begins in Africa.

Africa's identification as the centre where new forms of humans evolved depends on three quite different sources: biological patterns of endemism, fossils and genes. Genetics and fossil evidence converge in suggesting that the Modern human lineage began in Africa. They are also compatible in the timing, about 180,000–240,000 years ago. Since it is known that there were many and various types of humans already scattered over Africa and Eurasia at that time, there is a powerful corollary for the genetic evidence that many people have found hard to accept. This is that among all the individuals, families, tribes, subspecies or even species of humans, it was the matrilineal offspring of only one mother that gave rise to this new lineage. All the other genealogies eventually became extinct.

It is fortunate that an almost perfect skull survives to show what sort of people were about in Africa at that time. It is the Broken Hill skull (after which I named a biogeographic zone in the previous chapter). This human type has also been called 'Kabwe man' or, formerly, 'Rhodesian man'.

There are three or four other skulls, estimated to date from between 380,000 and 180,000 BP, found in places as far apart as Morocco and the Cape, to show that big-brained and big-browed humans were widespread in Africa. While they showed considerable differences (as could be expected over such a wide span of space and time), they have enough in common with each other to be a recognisably African type of human that differed from contemporaries in Europe and the Far East. The European and West Asian humans were Neanderthals while the similar but less well known Far Eastern people have been called Mapas. The African member of this trio is typified by the Broken Hill skull.

On the pagoda tree of human genealogy all three belong to the fourth tier, the Heidelberg radiation. The first branches near the trunk of any radiation are always more difficult to tell apart than their later, more divergent forms, and understanding human origins has been complicated by the superficial similarity of all robust early humans. Nonetheless, the African lineage *can* be distinguished and the recognition of its distinctness and continuity is important for two reasons. One is that it is almost certainly our own ancestral line, now worldwide. The other is that this polymorphic lineage has gone on throwing up primitive-looking types which have been confused with other, unrelated types of early humans.

The anatomy of the Broken Hill skull (and other related African fossils) represents a mosaic of primitive and advanced features. The latter anticipate those of contemporary humans (whereas both the non-Africans went off on their own anatomical tangents).

At first, Eve's offspring may have differed very little from other populations of the Broken Hill grade in their body build. Indeed, their differences could have been so tied in with psychological, cultural and other inconspicuous traits that it might be difficult, even impossible, for us to distinguish the earlier members of this lineage on the basis of fossil remains alone.

What is especially intriguing about the emergence of a genetic Eve in Africa is her coincidence with a distinctive 'Middle Stone Age' stone tool culture, which is first recognised in East Africa as early as 180,000–240,000 BP, in South Africa about 200,000 BP and in Ethiopia at about 180,000 BP. There is, therefore, a very suggestive correspondence between the theoretic birthday of a new human lineage and the estimated date for a more advanced and varied stone technology.

The sudden appearance of a new technology need not be matched by any strikingly novel form of human. Indeed, it is not necessary to expect that progressive changes in anatomy (notably, the thinning bones, lighter build and more 'childlike' heads that are typical of Modern humans) need always be associated with Middle Stone Age tools. If my basic hypothesis of 'self-making' is correct these changes could have been the *products* of a new technology and its associated behaviour. Should this be the case, gross anatomical changes could have trailed well behind the technical achievements of a genetically and intellectually unique population.

If the new technology and associated changes in the social system were the prime cause of further alterations in genetic and anatomical make-up, their influence was unlikely to be very fast. Furthermore, the physical results of that influence might have been quite irregular in appearance. This irregularity in the pattern of change and the trailing of biological adaptation *behind* technological innovation could go a long way towards explaining the variety of human 'races'. When locally adapted technologies dominated the existence of a distinct people, they became a major influence on the further evolution of that population.

Reduction and simplification of the teeth and enlargement of the brain are both discernible long-term traits in human evolution, but Middle Stone Age technology (and the presumably more complex social system that went with it) represented a change in tempo and intensity. Robust limbs and bodies, teeth and jaws could have been less useful when a larger group could cooperate more efficiently and bring in a bigger harvest than earlier small groups. Not only were powerful, heavily boned people at less of an advantage in the enlarged social system, but a more thorough artificial processing of food led to a similar decline in teeth, chewing muscles and their associated bony anchorage.

The position of the brain in relation to the face was originally determined by its being contained within a well-reinforced framework of thick bony struts. Only when bones thinned and the 'juvenile' growth period got longer and slower could the adult brain escape being contained within its bony cage. When this happened the ovoid cranium became the dominant volume in the head, the one to which all other structures eventually became subsidiary.

An increase in relative brain capacity implies that Broken Hill and other Heidelberg derivatives had made important advances on Erect humans. (Great variations in the total body or skull size of individuals make actual measures of cubic capacity rather misleading.) A plausible correlation of this was improved capabilities in speech. Thus, a crucial legacy from the Heidelberg/Broken Hill ancestry may have been a gift of the gab, as a social glue and means of sharing skills.

For foraging societies, the functional usefulness of a more developed language would have centred on better planning and coordination of all jobs linked with subsistence. Integral to planning in large groups is a greater differentiation of tasks for specific larger objectives. This would have been less a matter of permanent or fixed roles (which decrease flexibility) than the development of a wider range of structured activities, such as building traps, weaving long nets and organising battues or war parties. Indeed more fighting could have been one outcome of larger populations bred from improved efficiency and larger harvests. More frequent defence was, in turn, an incentive to develop better weapons. Acknowledged as being dependent on social organisation for their success, such activities would have been intelligently organised in order to increase the entire group's effectiveness as a sort of combine harvester. Introducing flexibility of role also implies a big increase in the complexity of languages that had to describe, discuss and direct the diverse activities that the group shared in. This may not have been an innovation in principle but it was a big advance in practical efficacy.

The feed-back from ever-increasing technical mastery into ever more complex language may have involved subtle anatomical changes. For example, sustained speech is thought to demand more refined control of diaphragm and ribs. Enlargement of nerve canals in the spinal column is one possible reflection of this. There is some confirmation for this idea in Dr Alan Walker's examination of Erects from Kenya where spinal nerve canals are measurably less developed than those of Moderns. Thus even the diameter of our nerve canals may ultimately be traceable to our development of technology!

As selection for physical robustness declined, being built of less material and using up less energy could have become advantageous. Two scientists, E. Trinkhaus and Y. Rak, have studied this in relation to gait. Rak has suggested that a realignment of the Modern's spinal column in its relation to the hip joints might have created a sort of shock-absorber effect that not only explains our economic bouncy walk but also contributes to the momentum propelling our bodies along with long easy strides. In this respect Eve's lineage had a decisive advantage over Neanderthals, which had less effective shock absorbers in the pelvis. Trinkhaus has described how the longer shafts of Moderns' legs are better suited to running than the stolid sustained trudging of Neanderthals. There are other subtle anatomical changes. Altered finger proportions favoured greater precision in the use of fingertips rather than a powerful grip by the whole fist. Modern shoulder blades are narrower and elbows less flexed, the entire arm is lighter and more mobile.

Larger, more cooperative groups may have encouraged individual survival. There is some evidence that Moderns lived longer than their predecessors. The benefits for shared knowledge (and more grannies to help out with food gathering and child care) are obvious.

I suggested in Chapter 1 that a coordinated pattern of attack and capture could have been a prime hominid advance over the techniques of other savanna predators. Any structured and consistent improvement in the efficiency of group hunting or foraging would handicap less well organised competitors. At first those competitors were carnivores but, with the passage of time, the main competition began to be between different hominid populations. The proliferation of different human types and regional demes would have been accompanied by significant differences in technical sophistication, if for no other reason than differences in the type, abundance, distribution and ecology of resources in different regions. The anthropologist Colin Groves was the first to emphasise the significance of discrete sub-populations of humans and to stress that they existed at each stage of human evolution. These would have been genetically distinct and it is on this mosaic of differences that natural selection would have worked.

Eve's emergence was a specifically local event in just one of an unknown number of regional populations belonging to the Broken Hill grade of humans. If Eve's lineage was initially little different from other Broken Hill types, is there any prospect of learning exactly where in Africa she originated? The question is not entirely academic, because knowing more about the broader ecological context of that emergence could throw light on key dimensions of prehistoric life, such as total home and day ranges and group size in relation to the land and resources that supported them.

So far the earliest unequivocally Modern human fossils are known from Tanzania, South Africa and southern Ethiopia – all estimated to date from between 100,000 and 130,000 BP. The fact that Modern humans were as widespread as this and also sufficiently common to turn up in the fossil record certainly implies that Eve's offspring had gone forth and multiplied.

The theoretical gap of some 100,000 years since first emergence is consistent with my contention that there was a lag as anatomical change trailed far behind technological, psychological, social, cultural and genetic changes. These complex and mostly inconspicuous characteristics would have defined Modern humans. While they slowly displaced, absorbed or eliminated earlier types of humans (probably in a staged, region-by-region progression), there may have been little to distinguish the victors from the vanquished in terms of crude anatomy.

If the origins of Modern humans are concealed behind rather unmodern bones, perhaps their technology may reveal more about beginnings. At present, the earliest direct association between fossils and Middle Stone Age tools is at the mouth of the Klasies River in South Africa. Dated to about 120,000 BP, the people here were extremely variable, some rather slight, others still very hefty. This polymorphism is consistent with a decline in the selection pressures that keep most wild animal populations within a narrow band of variation.

Other fossils demonstrate various intermediate conditions between Broken Hill man and recognisably Modern humans. A big angular skull from southern Ethiopia is called Kibbish (Omo I), and has been tentatively dated to about 130,000 BP. At the moment this is one of the most complete skulls (with a lot of the skeleton too) from this vitally important period in Africa. Dr Chris Stringer of the British Museum has shown that the measurements of this and other African skulls lie very close to later skulls from the Middle East of about 100,000 years ago and from Australian and European finds as young as 35,000 years ago.

In spite of a considerable time span, the fact that these diverse prehistoric people should have retained a basic similarity in the structure and proportions of their skulls strongly supports the idea that there was a relatively rapid diffusion of the Modern human stock out of Africa. In spite of heavy brows, receding foreheads and other conservative features, their general shape and proportions put them very close to the *centre* of variation for all Modern humans. These were not primitives in

Reconstructions of (left) Omo I 130,000 BP and (right) Ngaloba (Laetoli), 120,000 BP.

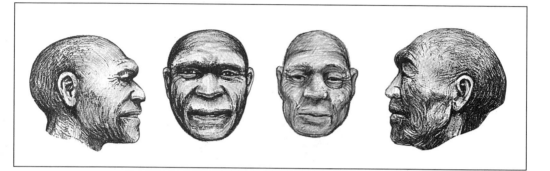

need of major transformations, rather, it was from variable people such as these that the 'races', as we know them now, would later emerge.

The range of variation that Modern humans are still capable of differs only marginally from that of these early fossils. Skulls from the last 100,000 years have confounded archaeologists because of huge variations in form. The main task is to distinguish between regional and adaptively meaningful details and the variability that is intrinsic to all Modern humans.

With a wide scatter of very varied fossils, often rather uncertainly dated, and an equally wide distribution of Middle Stone Age tools, are there any pointers for a centre from which the Modern lineage might have spread?

If a major effect of the Moderns' technology was to take a larger share of the resources – notably large mammals – this is less likely to have begun in an impoverished region with a poor or less diverse fauna. East Africa, as well as being equatorial, was also the most consistently well-endowed faunal zone. It was the area where the game was biggest but also where the competition would have been keenest. In less rich areas the rewards for greater efficiency could have been more marginal. My bets are on East Africa being the centre of dispersal.

Another indication might lie in the Moderns' prolonged distaste for cold latitudes. South Africa and the Ethiopian Rift can be very cool. Had Eve originated in either of these areas some useful cold tolerance and more rapid dispersion into temperate Eurasia might have been expected.

The question of climate is relevant to both the origins of Modern humans *in* Africa and their dispersal *out* of Africa. Temperatures progres-

Reconstructions of two individuals from Jebel Qafzeh (115 KY) and (91 KY). (left) Skull no. 6. (right) Skull no. 9. Even small samples show considerable individual variation in early Moderns.

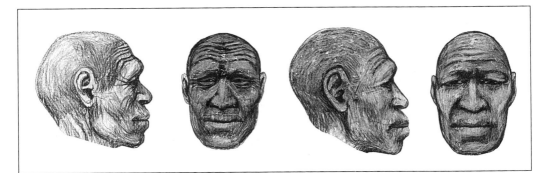

sively declined after 200,000 BP and reached an intense low at about 135,000 BP. Sea levels were at least 130 m lower at this time and it is possible that broader land bridges encouraged African Moderns to disperse widely. The climate then got warmer with startling rapidity about 130,000 BP. By that time Middle Stone Age cultures appear in southern Asia and the fragments of a Modern human at Zuttiyeh in the Middle East could suggest an arrival as early as 150,000 BP. However, the earliest reliably dated Moderns outside Africa are from Jebel Qafzeh in Israel, from some time between 91,000 and 115,000 BP.

Middle Stone Age sites in India are scattered all over the subcontinent, from near Amritsar and Jaipur to Hyderabad and Madras. They have not been found in either the mountainous or the more forested areas and are from what would have been fertile savanna lands not dissimilar to those of Africa. The assemblages of flake tools with many scrapers, points and awls are similar to those of the East African Middle Stone Age. The archaeologists Allchin, Goudie and Hedge, who excavated a Middle Stone Age workshop and other sites in central Rajasthan, found evidence for this industry flourishing during a moist period, and this is consistent with the prolonged (but interrupted) inter glacial between about 130,000 BP and 80,000 BP.

The Indian archaeologist V. N. Misra has pointed out that regional diversity eventually became quite marked during the Middle Stone Age, and he and his colleagues have made it clear that a significant population of Middle Stone Age manufacturers came to occupy fully all the Indian savannas, possibly from as early as 144,000 BP. If human fossils are ever discovered from this region and from this time span, it will be interesting

Reconstruction of skull no.5 from Mount Carmel and dated 80–110 KY.

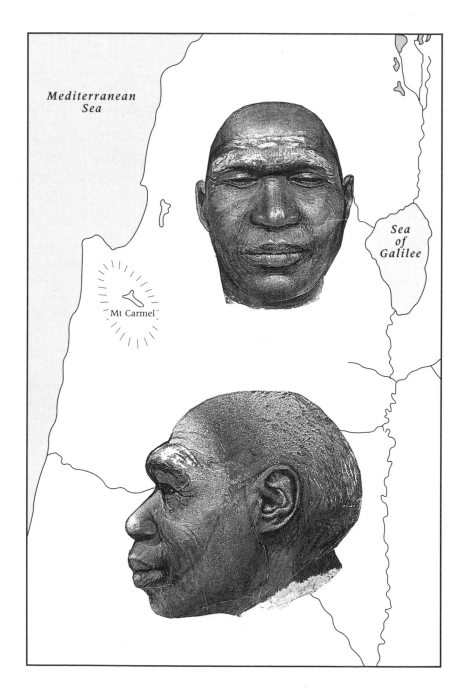

(overleaf) *Australiasia and possible paths of expansion followed by colonising population. Reconstructions from Keilor skull (right) and Cossack (left) show considerable variation in early Australians.*

to see if they too show a variety of robust and lightly built types such as those that were found at Klasies River in Africa.

The question of polymorphism becomes central to interpreting the next cluster of modern humans, so, before continuing with Asia, an excursion to Australia is necessary. Fossils undoubtedly related to modern Aborigines have been found at Keilor and Lake Mungo in New South Wales, where they have been dated between 38,000 and 25,000 BP. This is a big jump from the Middle Stone Age in India but there are indirect signs of human activity, especially massive firing of the landscape as early as 120,000 BP. The Mungo people were a wholly Modern type; indeed, one individual was strikingly thin-boned, slight and delicate. By contrast, much later skulls are very large, thick-boned, with heavy brow ridges and receding foreheads (which may have been exaggerated by artificial deformation in infancy or even by disease). The most extreme of these has the flattest and longest skull so far recorded. Found at Cossack in northwestern Australia, it is only 6,500 years old. In its general formation this male skull is wholly Modern and Aboriginal in character. If it is larger, longer and flatter than those of most people around today, a good diet in a favourable area and an individual constitution that was robust (in all senses of the word) are sufficient to explain his being somewhat beyond the norm. Massively thick Australian skullcaps may be the product of a pathological condition in which bone gets laid down at two or more times the normal thickness. Very strong contrasts can still be seen today, from big burly men in the Southwest to some really tiny people in northern Queensland and a lot of variety in between. Australian skulls that are earlier than Cossack but equally robust look less like modern Aborigines and more like generalised 'early Moderns'. This is the predictable drift from a highly variable but relatively generalised anatomy towards a more specifically regional type.

While the rest of the world showed a marked bias towards lightweights (a trend that has accelerated in the last 30,000 years), could there be something exceptional about Australia? Could there be ecological or cultural reasons for the retention of greater anatomical variability in Australia?

My explanation assumes that a very variable human genotype allows a relatively quick response to selection and that food and the environ-

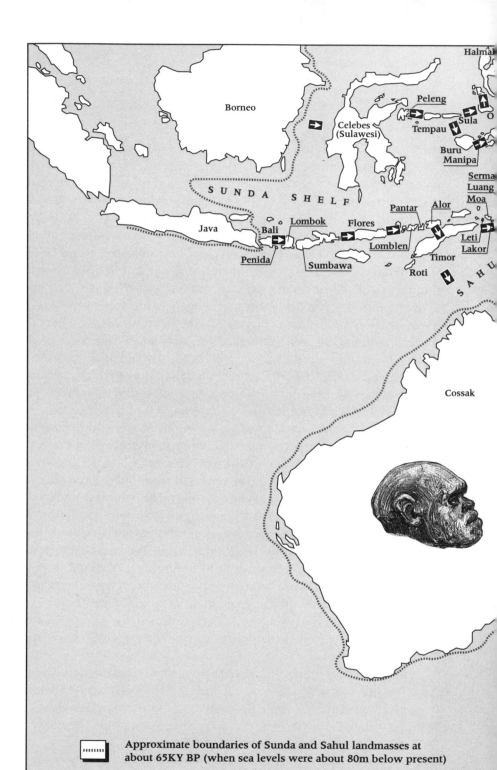

Approximate boundaries of Sunda and Sahul landmasses at about 65KY BP (when sea levels were about 80m below present)

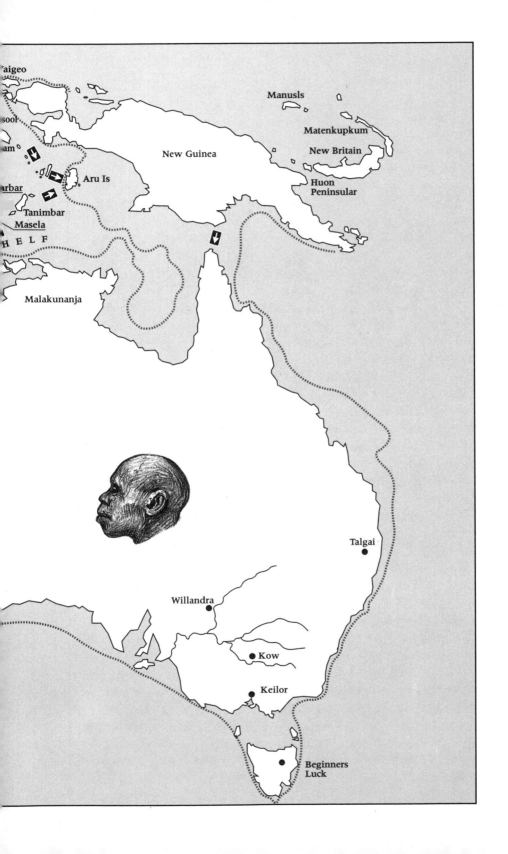

Waigeo

sool

am

rbar

Tanimbar
Masela

H E L F

Malakunanja

Manusls

Matenkupkum

New Guinea

New Britain

Aru Is

Huon
Peninsular

Talgai
●

Willandra
●

Kow
●

Keilor
●

Beginners
Luck
●

ment may amplify its effects. Since modern humans were originally robust, boniness in the body and skull can be considered the norm. There may have been an uneasy balance in Australia between opposing selective forces in favour of more or less robustness. If selection for the latter is tied in with relatively large, complex cooperative groups, these might have been impossible to maintain over large areas of Australia for very long periods of time. While climates were particularly dry and ecosystems correspondingly impoverished, the smallest foraging unit, a nuclear family, would have had many advantages by being self-reliant and its members robustly built. Not only is it known that climates have been harsh but there is also historical confirmation for this pattern of subsistence among some Aboriginal societies.

In a land that has no insurmountable natural barriers, Aborigines show a high degree of genetic and cultural fragmentation. This could be as significant as the early date of their isolation and sustained insulation from external influences. Physical demands on members of a very small family unit and greater environmental stress could have favoured body builds that are generally considered typical of an earlier stage in human evolution but may actually reflect a special type of environmental stress. If this explanation is correct it could throw light on many more problems than those of academic prehistory, because it demands greater awareness of the flexibility that is central to the evolutionary process and to human success as a highly adaptable animal.

In some fossils from Australia (notably those from Kow Swamp, Cohuna and Talgai), resemblances to Erect man, although superficial, have been sufficiently marked for some anthropologists to have proposed a direct line of descent for Australians from mainland Erect humans in eastern Asia! Numerous and detailed refutations long ago dispatched that thesis. Instead the robust Australians may actually offer an illustration of how durable and flexible the African legacy has been.

The earliest artefacts in Australia have been dated in the region of 60,000 BP and were found with pigments (and the grinding stones to prepare them) at a rock shelter called Malakunanja, below the Kakadu escarpment in northern Australia. The sea was about 80–90 m below the present level by about 64,000 BP and Malakunanja could have been over 200 km inland. First landing in Australia could have coincided with a period when the gaps between islands were at a minimum and the earliest low point was about 135,000 BP (but current evidence favours a later arrival). The sea dropped again about 109,000 BP and yet again at 87,000 BP and 64,000 BP. Each of these drops caused much of Indonesia to become a broad peninsula and joined Australia to New Guinea and the Aru Islands. The gaps between the Asian mainland (known as Sunda) and the Australian shelf (named Sahul) were more than 200 km at their narrowest and two main routes were possible for rafting colonists. The southern option went via Timor and the northern via the Moluccas. There are therefore various alternatives for ideal island-hopping conditions.

At present, all dated sites for Asian mainland artefacts bear much later dates than Malakunanja. Stone and bone tools from Long Rongrien on the Malay peninsula are only 37,000 BP while other artefacts from Japan, Da-phuk in Vietnam and Sai-vok in Thailand are even younger, about 25,000 years ago. By these times many outlying islands of the South Pacific had long been inhabited (New Ireland by 33,000 BP, the Solomons by 28,000 BP). The oldest artefacts from Papua New Guinea are waisted axes made of lava and recovered from Bobongara on the Huon Peninsula: they have been dated to more than 40,000 years ago. The deep interior of this giant island was well populated by 26,000 years ago, as is attested by numerous stone axes from Kosipe.

The Huon finds are of special significance, because this is one of the very few regions where the land is being steadily lifted and shorelines from all periods of the past are exposed. Papua New Guinea is the forward edge of the north-moving Indo-Australian continental plate and the Huon Peninsula represents an edge of that landmass that is actually being pushed up through its collision with the southwesterly drift of the submarine Pacific plate. The result is a staircase in which each step represents a former reef or shoreline. (Many are dotted with old shell middens.) Rises and falls in sea levels have failed to match the steady tectonic rise in Huon.

The last time sea levels reached or exceeded their present level was about 125,000 years ago; since then all previous shorelines, even the highest, have been submerged. Because low-lying tropical coasts were among the most favourable of habitats for early Moderns (especially during very cold, arid periods) all the premier occupation sites over virtually the whole of Southeast Asia are now submerged. This is the main reason for the scarcity of archaeological sites in a region where coasts provided much better living sites than the inland habitats (which were for the most part densely forested).

The beginnings of Modern human occupation in Southeast Asia must have depended upon an expansion east out of India. This might have begun at any time after India was effectively settled, but the fact that Southeast Asia is predominantly forested and frequently broken up into islands could have delayed colonisation. A single factor, climatic change, is likely to have precipitated or at least accelerated the advance of Modern humans into Southeast Asia. The glaciation of 135,000 BP is the earliest possible time for this expansion but it could have been substantially later.

Although the greater part of the archipelago is covered in dense forest, its flanks have a monsoonal climate that would have had a wider influence during cool arid cycles. Over a great loop running from Java and parts of Sumatra through Sulawesi to the Philippines, the dry season is severe enough to inhibit high rainforest. Today these areas are cultivated and support the main centres of population.

On the basis of his extensive knowledge of the region, the archaeologist Peter Bellwood has concluded that neither Erects nor pre-agricultural Moderns inhabited the rainforest core of Southeast Asia. Without doubting the existence of scattered coastal populations, he has suggested that the early hunting and food-gathering societies centred on these monsoonal islands.

At the time when Moderns first appeared on the scene, it is possible that Erects were still extant, at least on Java. If both Erect humans and a hostile environment were real deterrents to the inland settlement of Southeast Asia, it is possible that Modern humans had yet another pressure to become shoreline specialists. The ultimate outcome was, nonetheless, total replacement of the Erects at some unknown but possibly quite early time. What that confrontation might have entailed is open to question but how Moderns behaved towards other types of humans has probably been mimicked many times since, and by much more closely related types. There is no shortage of instances where people have driven one another out of entire territories but the duration, timing and mechanisms of such replacements in prehistory are still a mystery.

The question is particularly relevant for the Far East, especially China, where an abundance of fossils show that Erects were well established until as late as 150,000 BP. Whether Erects were totally or even widely displaced by the Oriental types of Heidelberg, Mapas, is another question. So far only known from a very few fossils, Mapa occurrences have been dated between 250,000 and 128,000 BP. If the career of their close relatives Neanderthals is any guide, it seems entirely possible that the penetration of inland China by Modern humans might have been substantially delayed because of a thorough pre-occupation of this region by Mapas and possibly by Erects as well.

It is in these circumstances that coast dwelling could have become a speciality of Moderns over a much wider area than the Indonesian archipelago. Once canoes and rafts were in general use Moderns would have been able to use the Chinese seaboard and offshore islands in a more thorough and productive way. From this fringing aquatic base they might have slowly pushed inland up the largest rivers and repeated (perhaps at a later date) the pattern of displacement that is so well documented for Europe. There Neanderthals collapsed before the advance of Moderns with startling rapidity in spite of the two types of humans having coexisted in the eastern Mediterranean for as much as 40,000 years. What is clear is that a very generalised type of Modern inhabited the shores and islands of the China Sea for a very long time.

The contemporary Ainu of Hokkaido could represent a northern very dilute relict of that first Far Eastern stock. There is a skull from the Zhoukoudian Upper Cave that has been variously dated from 11,000 to 40,000 BP and comes out closer in its measurements to modern Ainu than to any other living group. Others, of similar type are known from Longtan Shan and Liu Jiang. What is striking about these fossils and about the Ainu is the way in which they have kept some very generalised features that predate the divergence of late humans into well-marked regional types. The Ainu and their immediate prehistoric ancestors, the Jōmon, were more widespread in the recent past, and this physical type probably dominated the region around the Sea of Japan and the Yellow Sea for tens of thousands of years. Some of their characteristics are thought to be detectable in some Amerindian groups.

The specialised features of the Chinese, Eskimo and other Mongol peoples could have derived from a common ancestry with the Ainu. In

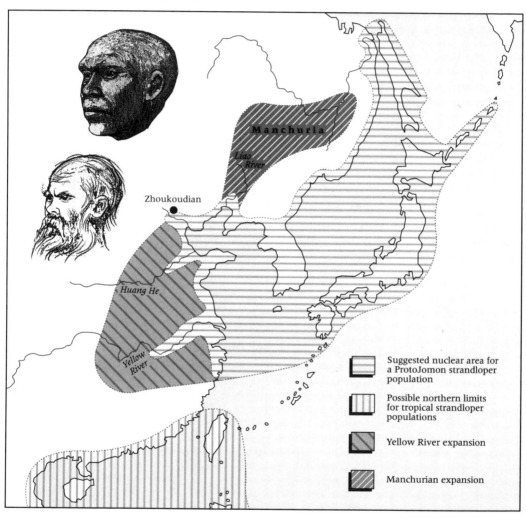

the absence of a fossil record, the most plausible explanation put forward so far is that just such a population penetrated inland (perhaps along the Liao River) and became established during the previous interglacial only to become trapped in Manchuria during the last Ice Age. According to this view, the 'late Mongo' physical type shows direct adaptation to extreme cold and their expansion has occurred during the last 17,000 years. In that short spell of time various Mongol populations are believed

Some post glacial archaeological sites in America. All are thought to be later than 16 KY and most are later than 11 KY. The extent of the ice sheet at about 18 KY is indicated.

to have crossed to the Americas, penetrated all of China, Southeast Asia and most of the central Pacific.

It is plausible that marine-based Ainu or Jōmon-like people might have spread around the Bering Sea and down the western coast of America during the earliest stages of the last glaciation. If they did so, their coastal camp sites are under the sea now, but there are Mexican and other sites that could be as early as 22,000 BP. (Still earlier claims have been made for at least two other sites in South and Central America.) If there was settlement during these early times, the people seem to have been both conservative and tentative in their exploitation of resources.

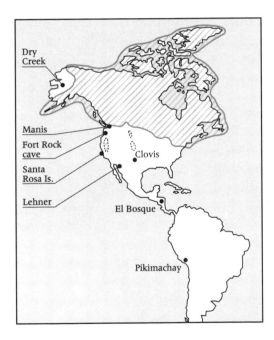

The main build-up of human populations and their spread inland began much later, with the first reliable dates at 16,000 BP and after. The more recent arrivals were undoubtedly of the later Mongol type but admixture with earlier Jōmon-like people might help explain the extraordinary variety of types that later developed in the Americas.

In Southeast Asia there is sparse but highly suggestive evidence for very widespread and early island-dwelling cultures, known to have been

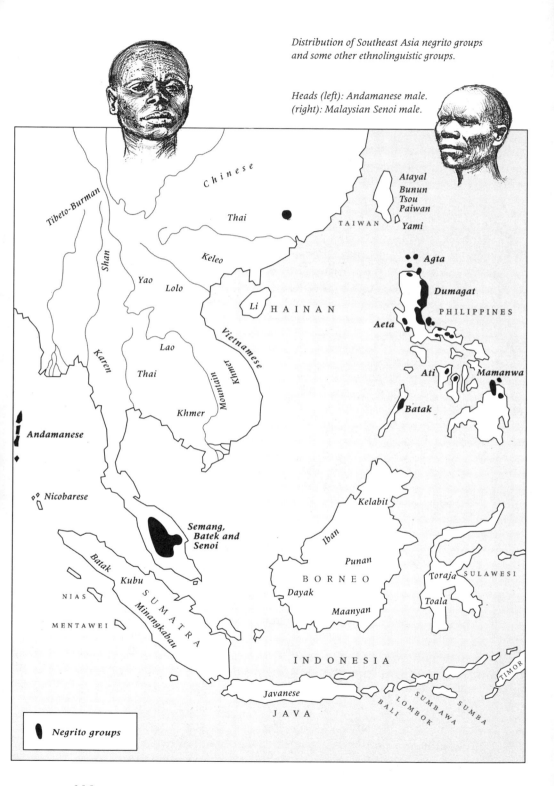

Distribution of Southeast Asia negrito groups
and some other ethnolinguistic groups.

Heads (left): Andamanese male.
(right): Malaysian Senoi male.

Chinese

Tibeto-Burman

Shan

Thai

TAIWAN

Atayal
Bunun
Tsou
Paiwan
Yami

Keleo

Yao

Lolo

Li HAINAN

Agta

Dumagat

PHILIPPINES

Vietnamese

Aeta

Lao

Karen

Thai

Mountain
Khmer

Khmer

Ati

Mamanwa

Batak

Andamanese

Nicobarese

Kelabit

Semang,
Batek and
Senoi

Iban

Punan

BORNEO

Batak

Kubu

Dayak

Toraja

SULAWESI

NIAS

SUMATRA

Maanyan

Toala

MENTAWEI

Minangkabau

INDONESIA

TIMOR

Javanese

BALI

SUMBAWA

SUMBA

JAVA

LOMBOK

● Negrito groups

112

present in New Guinea before 40,000 BP. The original focus for this tropical island culture would have been closer to the Asian mainland and therefore very much further to the northwest. The period in which sea levels rose and human populations became stranded on newly appearing islands, such as Java and Sumatra, would have been a very dynamic one. By 115,000 BP the outline of Southeast Asia resembled the shape it has today. Thereafter sea levels were mostly well below their present level and fluctuations went on continuously making, unmaking, joining and detaching the thousands of islands that dot the southwest Pacific.

The colonisation of Australia some time between about 120,000 and 60,000 BP is still the only indication of when seagoing watercraft might have been developed. The use of inshore boats and rafts would have preceded seagoing watercraft by thousands of years, particularly since the main centres of population must have been very far west of the continental shelves between Asia and Australia. If there is a gap between the time when terrestrial colonists first entered Southeast Asia and the time they plausibly started to use boats it is quite narrow, and boats could have begun to be used in the region not long after first arrival.

The natural boundaries of the Austronesian island world stretch in an unbroken chain from Sumatra to the Solomon Islands. This is a biogeographic unit that is definable through a diversity of plants and animals, such as screw pines, casuarinas, flying foxes and (in coastal waters) the dugong. For shore and island dwellers the chain continues round the Bay of Bengal, past the deltas of ten major rivers until the tip of India and Sri Lanka are reached.

In the middle of the Bay of Bengal, the 200 Andaman Islands and their black, spiral-haired inhabitants have long posed problems about origins and dispersal. The first British observers speculated that they were descended from Africans shipwrecked off a slaver's dhow. Later, Sir William Flower allied them with other black Southeast Asians and Pacific islanders. More recently, William Howells, in a series of detailed skull measurements, found Andamanese most resembled Africans but also found less immediate resemblances with Papuans and Australian Aborigines.

The slave-dhow thesis can be discounted, because the five Andamanese tribes spoke mutually incomprehensible dialects of a language reportedly unlike any other known. The implication of a remarkable cultural isolation of tribes within the Andamans is obvious. Preliminary excavations of middens have failed to confirm any ancient occupation of the islands and how long the people might have been there is still completely unknown. Flakes, blades and microliths have been found but are of uncertain age. Habitable caves with likely occupation deposits have been reported from Little Andaman Island but sustained archaeological excavations have yet to be made.

The apparent dual affinity of the Andamanese with other Southeast Asians and Africans is best explained by their being a westerly isolated relict of the Banda, who, I propose, originally inhabited virtually all the shores of the Indian Ocean perimeter. The lack of archaeological or genetic work on this important population is therefore a source of some frustration to all students of the origins of Modern humans.

Beyond the Andamans, up India's west coast, a fertile and well-watered shoreline with abundant seashore resources continues and archaeological sites south of the Indus delta show that littoral rivers close to the sea were occupied around 57,000 BP.

The movement or expansion of people over considerable distances is often imagined in individualistic terms, as if prehistoric groups were seized by the urge to explore or migrate. Such movements did not depend on individual wills; it was external events that imposed constant change and flux on human existence. A succession of bad years, incursions by aggressive neighbours, overpopulation, overhunting, the invention of a new and superior technique, fleeing from disease or fulfilling the prophecies of a shaman; all these and more could have triggered movement on and into the unknown. For a shoreline population resources were always richer in virgin territory and, along a coast, there is only one way to go – onwards.

Imagining whether prehistoric people perceived space as being empty or open is of course impossible, but it is not unrealistic to extrapolate from more recent history and recognise that 'new frontiers' are very much a matter of matching up techniques to resources. Efficient small-game and plant gatherers view the vast territories of mega-hunters as 'wasted'. Colonising farmers of every description have always viewed the land of foragers as 'unused' and moved in on it. In this way it is easier to understand how one population repeatedly replaced another, and how a distinctive group, such as shoreline strandlopers, would have found little opposition to their expansion along tropical shores, which they were uniquely equipped to exploit by virtue of their boats and well-tried harvesting techniques.

For people who had developed rafts and boats there were no physical barriers to their settlement of virtually the entire seaboard of continental Asia and the Indian Ocean shores of Africa. During arid periods, the drier coasts may have invited occupation only during the wet seasons or at spots close to sources of fresh water (parts of south Arabia and Somalia being the most forbidding stretches in this respect). Nonetheless, wherever there were hills or mountains in the hinterland that were sufficiently high to catch mist or rain off the Indian Ocean, they generated springs, streams or rivers, all potential foci for settlement. Even the more barren coasts held temptations for seashore gatherers. Both the south Arabian and Somali coasts are still major turtle hatcheries and the beaches abound with molluscs, crabs and various seasonal gluts of marine small fry. Offshore are some of the richest fishing grounds in the entire ocean. The hinterland of these coasts would also have offered some subsistence before they were degraded by overgrazing and timber felling.

The known history of Socotra Island provides some insight into the changes that have taken place along the coasts of the Arabian Sea. Ptolemy's voyage there, nearly 2,000 years ago, reported flowing rivers, crocodiles and giant lizards. In 1612, the island was still sufficiently moist to support water buffaloes but the present stock comprises camels, sheep and goats and there is no surface water outside the wet season.

So far, archaeological sites for the Middle Stone Age have not been found on the tropical African coasts of the Indian Ocean. However, along the temperate shores of South Africa, marine economies had emerged from a local indigenous base. These are known from sites that remain close to the sea today because the continental shelf is steep off the eastern Cape and Natal. Further north, seaside camps are under water and out to sea, so there is little chance of verifying the arrival of a new culture. In spite of this, there are signs of a new mode of subsistence, albeit one that had already accommodated to living inland.

Between the Limpopo and the Zambezi rivers and along their main tributaries a distinctive family of stone tools, known as the Charaman, consists of flake blades, core axes, picks and small bifaces, which were more likely to have been used for working wood and other plant material

Map of likely Charaman sites in southern Africa.

(and possibly for digging up roots) than for processing animals. This is in contrast to earlier Middle Stone Age tools from southern Africa, which were largely applied to animal matter. Today, the dominant vegetation in southeast Africa is fire-induced and modified by humans but the region still has many common nutritious and highly productive food plants.

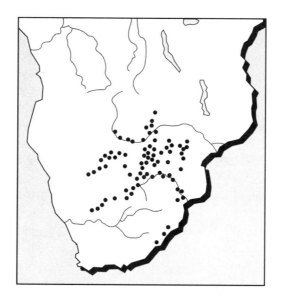

Charaman artefacts are of uncertain age but may be more than 50,000 years old. Could the Charaman reflect the arrival of a new population of colonists moving inland from the coast? Down northeastern and much of eastern Africa the hinterland behind the coastal strip is arid. Parts of Mozambique and most especially the mouths of the Zambezi and Limpopo rivers are rich, well-watered lands, where a scattering of seashore fisher-gatherers could use these rivers to expand inland. Here, there would have been fewer inhibitions for a steadily growing population to develop an array of tools and equipment suited to a new, non-marine existence that exploited a much wider range of small-sized foods (both animal and vegetable) than did large mammal hunters.

One indication that this relatively sudden eruption of a new culture had an external and eastern origin is its interpolation into the older and very widespread Middle Stone Age culture. Still predominantly oriented towards large animal hunting in savanna, this tradition had begun to acquire some regional peculiarities about the time the Charaman appeared. A decidedly tropical origin for the Charaman is suggested by their preference for warm lowlands. They left the cooler uplands to the big-game hunters. Instead, they followed up river valleys as far as the Okavango delta, Lake Ngami and the Malopo. They penetrated deeply into a savanna industry that was formerly known by the generic term 'Stillbay' (but had many regional variants). Stillbay-like industries were characterised by lance-shaped spearheads, hand axes, scrapers and cleavers with carefully chipped working edges. They dominated drier habitats from Ethiopia and the Horn to the Cape.

The interpenetration of two distinct economies in southern Africa gave rise to successors that seem to have been shaped by both previous cultures. Nonetheless, an ecological and regional division developed that followed a new and somewhat different pattern. All African cultures diversified after this point but some older traces have remained recognisable. For example, the precolonial culture of the Khoisan conserved a few elements that resembled the very ancient Stillbay traditions.

Charaman-like techniques were later refined to a very high degree in the Zaïre basin by a culture known as Lupemban. These people developed a technique of using a punch to flake off long, sharp blades of stone, which were then retouched with further very fine and controlled chipping. Their long, pointed stone spearheads are among the finest stone tools ever made. Double-ended stone picks and heavier, pointed picks suggest that more sedentary living and digging may have become important.

The Lupemban was predominantly a tropical, riverine and forest culture that insinuated its way along moist river valleys and lakeshores around Lake Victoria and over much of tropical west Africa. There was, therefore, a very interesting dispersion of at least two contrasting Middle Stone Age cultures across Africa. The newer industries eventually came to dominate all the moister regions of southeastern and central Africa, as well as the westerly equatorial zones where higher rainfall generated an abundance of rivers and a more heavily wooded and forested country. It focused on the more numerous, more diverse but smaller-scale fodder of humid areas. By contrast, the older established tradition may have relied upon a less intensive use of resources with a strong preference for

Three contemporary Africans:
Left – Mangbetu woman, Zaïre.
Centre – Fulani or Fulbe man, Guinea.
Right – Mtumbi man, Southeast Tanzania.

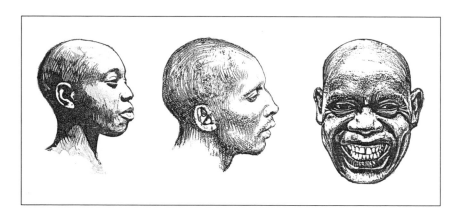

animal foods. The people were more nomadic in their pursuit of large herbivores, which flourish best in the drier, more open savannas of the east and south.

About 45,000 years ago, populations in Africa began to increase very substantially and with that increase came a proliferation of cultures. In the Zaïre basin, the Lupemban elaborated still finer, more complex microlithic techniques and gave rise to an industry known as Tshitolian. In Zambia, another distinct type is called Nachikufan, while yet another branch developed in west Africa; one of its more recent manifestations is named after the Ghanaian village of Kintampo. With these developments the Stone Age begins to approach the immediate origins of contemporary people.

Above the equatorial belt in northwest Africa, the artefacts of an industry known as the Aterian appeared for the first time while the Lupemban flourished further south. This became very widespread in the Sahara and lasted until at least 24,000 BP. Aterian spearheads, points and knives are highly distinctive; manufactured by pressure flaking, they typically have well-shaped hafts for binding and gluing into wooden shafts or handles. The Aterian clearly continued the large-animal hunting traditions of the earlier Middle Stone Age peoples of arid and semi-arid Africa. The Aterians gave rise to several regional variants, especially along the Mediterranean coast, notably the Capsian industries.

The Capsian was a localised North African industry which, at one time, was thought to have provided the foundation for the Aurignacian

(overleaf) Chart of physical expansion and
cultural events in Old World and Australia
from 180 KY.

ADAM ABROAD

invasion of Europe. Although it is clear that an originally African and
Middle Eastern population did expand into Europe, it was not related
directly to the Capsians. In fact, the expansion into Europe was earlier
and its protagonists were contemporaries of the Aterians. With that
event it could be said that a new phase of cultural and genetic diversification began.

This is a convenient point to take stock.

Before 45,000 BP, humanity formed fewer regional subdivisions than
it does today and the main blocs may have fallen into the following broad
categories. There were direct descendants of the first Moderns still
occupying the African savannas. These formed northern and southern
populations separated by a broad equatorial band of people that would
have had a measure of the same ancestry but were also partially (perhaps
predominantly) derived from the Southeast Asian immigrant Banda.

Very early expansions out of Africa would have populated the warmer
southern regions of Asia. Those in the more westerly parts were
beginning to advance north by 45,000 BP and would later become the
Europeans.

Further east, in India, savanna hunters maintained traditions similar
to those in Africa but the moist coasts and the south would have
belonged to a different diaspora. Sri Lanka and the equatorial forested tip
of India would have been inhabited by outliers of a scattered Southeast
Asian complex.

From small beginnings as a marginal continental group, the very
distinctive Banda had expanded to populate one of the most extensive
areas of settlement anywhere in the world. Because their realm was so
extensive and fragmented, the Banda would have been very diverse
genetically. Their isolation, however, would have been partially offset by
the mobility that their watercraft gave them, so that migration and
mixing would have been continuous. It was a very early offshoot of this
population that first reached Australia. Although later influences clearly
filtered down from the north, most of the Australian population has
probably derived a large part of its inheritance from the earliest landings.
Sri Lanka (Ceylon) might have shared in a similar early settlement.

Of the very first Moderns to arrive on the shores of the China Sea
nothing is known but a few rather late Ainu-like skulls. Where the
Mongols originated is still uncertain, but Manchuria and the Yangtse

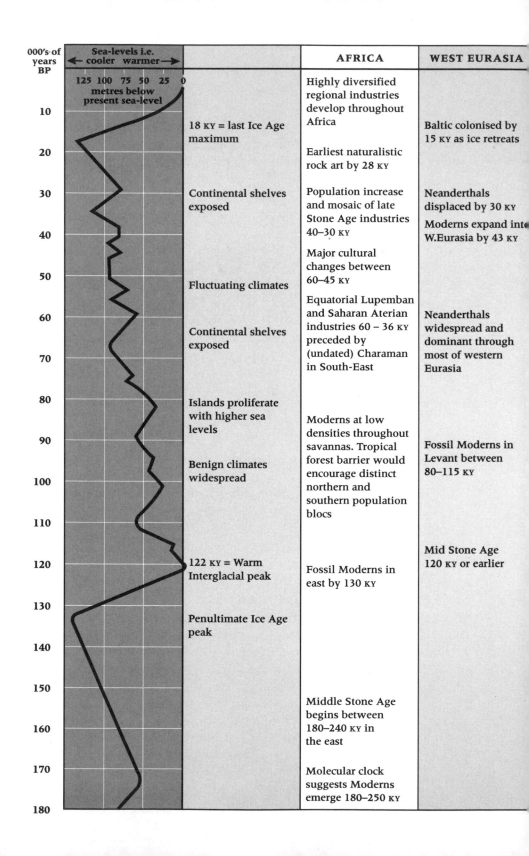

000's of years BP	Sea-levels i.e. ← cooler warmer →		AFRICA	WEST EURASIA
	125 100 75 50 25 0 metres below present sea-level		Highly diversified regional industries develop throughout Africa	
10				Baltic colonised by 15 KY as ice retreats
20		18 KY = last Ice Age maximum	Earliest naturalistic rock art by 28 KY	
30		Continental shelves exposed	Population increase and mosaic of late Stone Age industries 40–30 KY	Neanderthals displaced by 30 KY
40				Moderns expand into W.Eurasia by 43 KY
50		Fluctuating climates	Major cultural changes between 60–45 KY	
60		Continental shelves exposed	Equatorial Lupemban and Saharan Aterian industries 60 – 36 KY preceded by (undated) Charaman in South-East	Neanderthals widespread and dominant through most of western Eurasia
70				
80		Islands proliferate with higher sea levels		
90			Moderns at low densities throughout savannas. Tropical forest barrier would encourage distinct northern and southern population blocs	Fossil Moderns in Levant between 80–115 KY
100		Benign climates widespread		
110				
120		122 KY = Warm Interglacial peak	Fossil Moderns in east by 130 KY	Mid Stone Age 120 KY or earlier
130		Penultimate Ice Age peak		
140				
150			Middle Stone Age begins between 180–240 KY in the east	
160				
170			Molecular clock suggests Moderns emerge 180–250 KY	
180				

INDIA	EAST ASIA	SE ASIA	AUSTRALASIA	000's of years BP
	Millet and Rice cultivated in China by 8 KY	Foraging economies give way to agriculture	Outer Pacific is colonised by Pelagic Polynesians	10
	Pottery by 13 KY Bering Straits crossed by 15 KY	Agriculture by 15 KY preceded by long period of 'proto-horticulture'		
ate Stone Age by 2 KY			Axe culture in New Guinea Highlands by 25 KY	20
		Sahul land-mass extensive 34 KY Small flake industries widespread	Arrival in Solomon Islands by 32 KY	30
				40
oastal industry 0–57 KY N.W.India	Uncertainly dated Modern in China by 67 KY	Higher sea-levels isolate numerous marginal communities	Confirmed arrival by 52 KY	50
			Possible arrival by 70–58 KY	60
				70
				80
Watercraft allow coastal expansion by strandloper economies along Indo-Pacific shorelines (dates unknown)				90
				100
	Mapas may be dominant in East Asian interior	Arrival of Moderns most likely 120–90 KY (Many islands) Last Erects estimated at 100 KY	Remotely possible first arrival	110
Mid Stone Age egins between 20–163 KY (144 KY?)				120
				130
				140
				150
				160
				170
				180

Suggested sequence of primary dispersal by Modern humans.

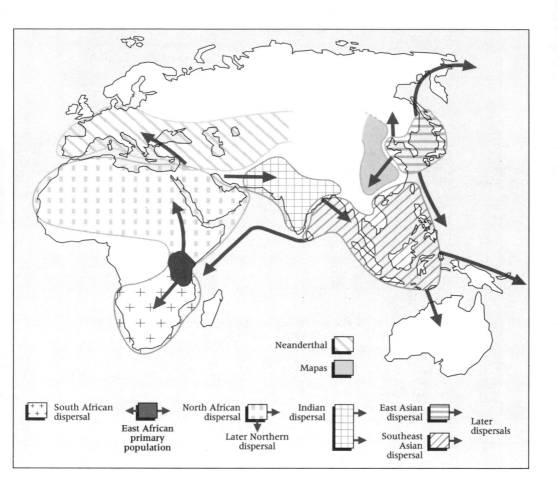

basin are the two main possibilities. In either case, the Mongols made the fastest and most far-reaching eruption of any prehistoric people, penetrating to the end of South America, overrunning much of the former Banda homeland and initiating further waves of colonisation and exploration in the Pacific.

It has not been my intention to propose a comprehensive history of human dispersal and diversification. Rather, I wanted to present the pattern that has emerged from a biologist's examination of a scatter of fragmentary and apparently disconnected clues. If the course I sketch out in these chapters seems long on speculation and short on facts, I have *(left) An analysis of skull shapes by W. Howells. (right) A reinterpretation of Howells' tree as the product of a radiation influenced by migration and genetic mixing.*

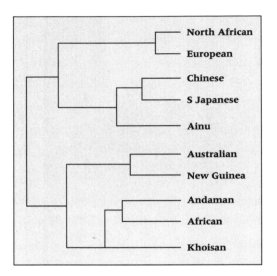
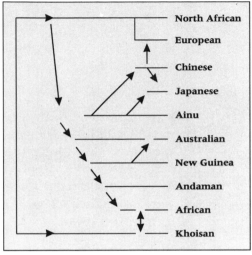

offered this interpretation in the knowledge that most of the alternative scenarios are even more lacking in factual grounding and their biological credibility is shorter still.

I therefore offer this and the following chapters as a personal exploration and a working hypothesis for human dispersal. It is written in the conviction that no portrait of any part of prehistory can claim much depth without its biological dimensions.

4: From Nuts and Mega-Meat to Clams and Yams

It is common to find animals identified by their food – anteaters, fig parrots, root rats. An alternative name labels a mode of living – kingfisher, nutcracker, oystercatcher, harvester ant. Formal taxonomy also uses consumer titles, such as Carnivores and Insectivores, or classifies with names that describe a way of catching or processing food – divers, flower-peckers, ruminants. So food and how it is got are seen as fundamental to what animals are in both everyday and scientific language.

Food and food-getting are central to the study of ecology and are certainly no less relevant to human prehistory and understanding what humans are. What sort of eaters humans have been has depended very largely on what sort of catchers they have been. A great part of human prehistory concerns change, innovation and diversification in food-getting techniques, but first came the challenges that wild animals and plants presented to the ingenuity of early humans.

A huge range of potential foods surrounded early humans wherever they were but the greatest variety and quantity was in the tropics. Nonetheless, in any one locality or at any particular time the list would have been far shorter. Many animals and plants were available only at certain times or in certain areas and even then the smallest sample reveals a barrage of difficulties, obstacles and protective devices that had to be overcome before the potentially edible could become an actual meal. The greatest narrowing of choice would have derived from strongly conservative eating habits and these would have been reinforced and maintained by inherited tool kits and the techniques that went with them. Notwithstanding their abundance, the greater part of the potential foods awaited the invention of special techniques to make them available or accessible.

The expansions of diet that are described in the following pages assume that innovations in both foods and food-gathering techniques tended to be tied in with a steady expansion of range and habitats. A

An Agta family, Baclai, Philippines (Photo J. Kamminga).

single African lineage spread into new lands and encountered other human (but non-Modern) populations. In their responses to these challenges and to fluctuating climates they changed diets, changed physically and laid the foundations for a variety of food-getting cultures, many of which are still with us today, albeit in much-modified forms.

Since his wanderings are a central theme of the 'Self-Made Man' they provide a series of steps which can be tied in with gross differences in diet. For example, terrestrial foraging can be polarised into the hunting of large animals at one extreme and the gathering of plants and small animals at the other. By contrast the water's edge offers the plunge into almost wholly aquatic fishing on the one side and, on the other, beachcombing for the animals and plants peculiar to the littoral.

Any one broad category offers scope for further specialisations in diet – savanna megafauna versus forest diversity, mainly animals or mainly plants, island fauna and flora as opposed to mainland ones, mobile food versus sessile food, fish versus mammals, roots versus fruits and so on. Such simple categories were certainly defied by the realities of prehis-

toric life but they do help throw the early differentiation of human societies and their diets into stronger relief.

The model that follows draws from many indirect sources because the physical evidence is very largely lost for ever. The remains of prehistoric meals are seldom preserved and the bits that do fossilise present a very skewed picture of early human subsistence.

There are great areas of the globe where stone tools are all that can be found of past human activity; in other localities there are some human bones; rarer still are middens where hard parts of prey, bones, teeth and shells give a short list of animal foods. Rarest of all are the remains of plants, yet in some regions, particularly the wet tropics, plants must have been the main staples. Some of the most nutritious species entered domestication long ago and are now a familiar sight on market stalls, although in a very altered form. The properties that made them successful candidates for the human larder become clearer as their original ecological niches are investigated.

The great majority of wild foods are now relatively rare or localised because the richest lands, where prehistoric humans would have fared best, are now cultivated or ranched. Hunters and food-gatherers were pushed into marginal or inaccessible regions long ago, so the spectrum of species they exploit is generally smaller than for many prehistoric communities. Nonetheless, the food ecology of contemporary and recent hunter-gatherers can be some guide to the past (always allowing for the fact that most of them have acquired modern aids such as dogs, improved vessels, nets, projectiles and sometimes even guns). Innovations, such as hunting with dogs, immediately alter the choice of prey, because the hunters' energies will be concentrated on those species that are most vulnerable to dogs. Most specialised hunting devices tend to focus efforts in similar ways.

Survival in the more distant past involved fewer aids but greater strategic versatility in the use of resources. Extinction of so many large species means that the choice and abundance of large prey would have been very much greater than for any living hunters. There are four distinct but interrelated influences that would have made feeding ecologies differ very significantly from region to region.

Firstly humans have inbuilt nutritional needs and physiological limitations that must be met. These influence both the choice of foods and the situations in which they can be collected. Some foods and some places need adapting to and, once achieved, such specialisation can become very specific and self-perpetuating.

Secondly, the particular choices of food that would have been available in different parts of their range would have influenced ways of life and types of technology and here, too, established habits could have encouraged conservatism in diet and economy.

Thirdly, there would have been constraints in terms of defences put up by the plants and animals and limitations imposed by the places in which they were found. Diseases, predators, nuisances and direct competition from other animals would also have helped to define and maintain what foods could or could not be eaten in any one region.

Fourthly, technology was the key giving access to otherwise closed food supplies. Technology also helped to overcome some of the constraints.

For each of these influences there are enough clues, direct and indirect, to allow tentative reconstructions to be made of regionally different diets in prehistory. Each factor in turn would have had the potential for equilibrium and stability or for change, variety and innovation. When archaeological and other evidence is examined for the operation of these four influences and for signs of stability or change interesting patterns emerge.

The first Modern humans inherited both an environment and a food supply that would have differed only in quite minor ways from that of their less evolved predecessors. Their primary innovations probably concerned things that have not been preserved in the fossil record, such as improvements in communication that made for more efficient co-ordination and organisation within the group, larger social units and perhaps higher overall population densities.

The well-tried, long-established tradition of killing animals probably continued but more intensively. While the mega-hunter would have eaten many plants, particularly nuts, fruit and seeds, his primary view of vegetation would have been that it was food for game. The various grazing, browsing and fruit-eating habits of animals, what they preferred and when would have been well known. In ecosystems where animals were exceptionally abundant, as over much of the East African savannas, the plant foods that humans could eat would have been subject to direct

A hunting camp exploiting large mammal migrations in the south Sudan. Smoking racks to preserve meat are surrounded by grass sleeping huts (Photo J. Kamminga).

competition from other animals, such as primates, pigs, elephants and birds.

For some of these foods tools gave humans special advantages, such as fire and stone axes to get at honey, picks for buried roots and at some stage or other the neutralising of toxins by soaking and cooking. Nonetheless, the existence and anatomy of Nut-cracker men and other specialised vegetarian primates show that teeth, jaws and temperament are essential parts of being vegetarian. *Their* specialisations in this respect serve to emphasise their absence in *our* lineage. Neither Erect nor early Modern humans competed intensively with vegetarians because that was probably the main preference of their australopithecine cousins. When viewed as food, large animals are the most concentrated and bulky packages of nutrient about. With few other predators able to penetrate their defences, it was the larger animals that offered the most cost-effective targets for the already very effective hunting technology of African *Homo*.

However, such animals were not mere food on the hoof. Plant communities had co-evolved with large herbivores over many millions of years and the impact of their feeding was often a force comparable with that of the climate. Until recently, the passage of a large herd of elephants could be likened to the mess left behind by a hurricane.

Upuk, a ten-year-old Agta boy with a pig he has killed (Photo J. Kamminga).

Whether plants were eaten or avoided was part of an evolutionary struggle between eater and eaten which dominated the prehistoric plantscape and would have determined where, when or whether many a plant or plant association flourished or declined.

In Africa, India and many parts of Eurasia, herds of large herbivores belonging to twenty or more species traversed every habitat that could support them. Their pathways linked pastures and waterways as prominently as roads link modern towns and villages. They were as decisively architects of the landscape as people are today, and it is certain that the ecological influence of humans would have remained subsidiary to that of large herbivores until the numbers and variety of the latter had been very substantially reduced. Such a dramatic usurpation of ecological dominance was achieved by Modern humans continuing and intensifying a basic trait of their lineage – which was to be the super-predators of large mammals.

It is for this reason that small bands of modern hunter-gatherers living in impoverished animal communities and on lands that are marginal for other humans are often inappropriate models for early hunting subsistence. Anthropologists have sometimes seized upon those occasions when they could record the killing and consumption of a large beast, but such events are now very rare and certainly never routine. Indeed, so

129

remote are large mammals from the normal diet of the Anbarra Aborigines in Arnhemland that Betty Meehan, an Australian anthropologist, has described a hunter, Bandarrpi, shooting a buffalo with his rifle and then leaving the carcass to rot. Preferring shellfish and turtles, his family and friends were uninterested in butchering it.

The decline of large mammals is relevant to prehistoric diets at two rather different levels. Their direct relevance is that in their heyday they were the chief food resource for humans. The other concerns a larger reconstruction of prehistoric ecosystems. It is certain that the passing of large mammals has altered the scene in ways that have not begun to be measured or assessed. A pioneering study by Dan Janzen in Costa Rica has suggested that the life cycle of many long-lived plants was perturbed by the loss of large browsers and dispersers. For example, he has identified a limited capacity for dispersal in certain giant-seeded trees. There are other clues to some of the ways in which ecosystems dominated by giant herbivores were different from today's.

The first implication of their abundance is that a great part of fresh growth – grasses, leaves, twigs or fruit – would have passed through their guts. This would have been excreted to provide a rich source of nutrients. The immediate beneficiaries were normally insects, but at times of acute shortage, semidigested cellulose provided emergency rations for various other herbivores. Several species of giant dung beetle remain in those parts of Africa where there are still some elephants and rhinos. They are capable of burying great quantities of fresh dung in a matter of hours. Their larvae are then greatly sought after and dug up (notably by aardvarks and honey badgers). Older, drier dung is consumed by many species of termites, which in turn are food for many insectivores, bats, birds and lizards.

The inevitable by-products of massive consumption would have been extensive damage, waste and trampling as well as final excretion of digested cellulose – a lot of dead plant material about. In drier areas, where plant growth was strongly seasonal and giant consumers were at high densities, the secondary cellulose eaters, such as termites and beetle larvae, would have so cleaned up land surfaces that fires would have been unable to take hold for lack of fuel. In other areas, such as sumplands where herbivores were seasonal visitors, fires might have had more fuel and persistent damage to trees by elephants and rhinos would have decreased their fire resistance and therefore helped to promote very open, grassy valleys where short grass regrowth would help support smaller grazers. Trampling would also have encouraged repeated grass and herb flushes so that many specialised 'fresh-growth-grazers' and 'new-leaf-browsers' would have followed in the wake of the herds. Many other species could have benefited from picking up 'waste'; fruit, stems and roots could have fed monkeys, birds and pigs.

Gigantic consumption could be sustained in some habitats and at some seasons, and big mammals would have concentrated in mega-herds only at certain times, drawn by a major resource or through necessity, as when water was short. The physical effects of concentration around or near seasonal water sources can hardly be exaggerated. In general, large mammals open up vegetation through browsing and grazing but the long-term effects of concentration can be seen at waterholes in several modern African national parks. Open arenas form close to the water with wide, well-trodden avenues leading out in every direction. Persistent trampling creates hardpans surrounded by a few plant species that are quick to repair damage or are distasteful or poisonous to elephants and rhinos.

Chemical defences would have been widespread among plants in giant-dominated savannas; other deterrents included thorns, hooks, coarse, well-reinforced barks or rubbery, flexible and very tough branches. Palatable species specialised in rapid surface growth and regeneration in hidden root stores. Most of these defences would also have deterred humans.

Much of the landscape would have been dominated by giant mammals but they would have avoided rough, steep hillsides and rocky outcrops which became preferred sites for human shelters, factories and resting-up areas. Grasslands, such as those on quick-drying screes, might attract no more than brief seasonal interest from large mammals because growth was so ephemeral. Mature climax forest would have put most of the vegetation out of reach, so that forest zones would have supported fewer large mammals and then only along water courses and in openings. Really wet tropical forests were equally unattractive to humans.

The activity of large animals creates physical advantages for humans, and it is no coincidence, perhaps, that opened-up country appeals to people even today. Numerous smooth pathways and open ground would have allowed for faster and safer movement, especially for a bipedal, hoofless animal.

Humans would have made way for very large herds and might have avoided overcrowded concentration grounds, but their knowledge of and proximity to the animals themselves would have been intimate and continuous. Pre-Modern humans might have put that knowledge to practical use. Modern humans continued that tradition and, so long as large mammals continued to be abundant and accessible, there was no incentive to abandon mega-meat as a major staple in the diet. The human digestive system is able to maintain an almost wholly carnivorous diet but it can be surmised that seasonal gluts of fruit, seeds, nuts and (rather later) roots were also harvested by early Modern humans and that they generally had a diverse diet.

For many of these foods there might have been competition from other animals but nuts, such as those of the Manketti tree (*Ricinodendron rautenenii*) or marulas (*Sclerocarya*), were probably important human staples. Here were rich sources of carbohydrates, fats and protein with the added advantages of being storable and impervious to the attacks of insects and competing primates. Both large mammals and hard nuts were resources that could be released only through the use of tools and artefacts.

The African savanna ecosystem is remarkable in its enormous extent. An ecological community that included many fire-resistant trees and grasses once ranged from one end of Africa to the other, from north to south, east to west and over lowlands, hills and high plateaus. Forest and woodland intruded in the west and centre, desert steppes in the southwest and north; but for the rest a single complex of wooded grasslands could be characterised by its fauna and flora – the elephants, antelopes, acacias and grasses that typify today's great savanna parks from Senegal to South Africa. The only other region with similar habitats and fauna is India, and the savannas there once supported a comparable but smaller range of species.

The savannas of India not only resembled those of Africa, they must have been exploited by humans in similar fashion, because stone tools from the Acheulean up to the Middle Palaeolithic are almost indistinguishable between the two regions. The resemblance between the two regions is best appreciated by comparing some of the common animals that were most likely to have been eaten regularly. Large mammals come top on the menu. The intervening region of Arabia used to have a richer fauna than today and resembled a short list of North African species. Actual bones found on Stone Age sites in India include species of elephants, rhinoceroses, wild horses, giraffes, extinct giant cattle, buffaloes, deer, bluebuck, antelopes, gazelles, wild sheep and goats, pigs, porcupines, monkeys, bears, ostriches, crocodiles, turtles and freshwater molluscs.

For evidence of Modern humans' expansion eastwards out of India, there is, as yet, no guiding archaeological record but there were two main ways to go. Across the hills and valleys of Burma and Yunnan were the rich plains of China. Here there could have been some continuity of fauna and diet. Several species of fossil elephants, rhinoceroses, horses, cattle, deer and pigs figure prominently in northern Chinese cave deposits that are more than 150,000 years old. There is nothing to indicate any break in large mammal hunting traditions in Far Eastern Asia but neither is there any evidence for the early arrival of Moderns, and other human types held sway in China very much later than in Africa.

The second dispersal route out of India swung to the southeast. In contrast to a mosaic of steppes, savannas and forests in central India, the natural climax of growth for the humid Indonesian archipelago was forest. Local influences such as poor soils, drier winds and rain shadows would have diversified habitats but large mammals would generally have been scarcer. Thus Moderns would have shifted into an entirely different ecological setting as they travelled down the eastern shore of the Bay of Bengal.

There is a big gap in the archaeological record for this region and for the period of initial invasion, but humans inhabited the Bornean cave of Niah for well over 30,000 years. This immense cave extends over ten hectares and reveals that pigs, monkeys, orang-utans and a giant pangolin, all decisively forest species, were the commonest terrestrial prey. The cave is about 16 km from the sea but changing sea levels would have made the shore at times nearer and at times further away. The remains of aquatic prey include fish, reptiles and molluscs.

In the regions and periods that have been discussed so far, Modern humans not only displaced but took over the large mammal hunting economies of their predecessors. In Southeast Asia, they entered a zone that was not only very different ecologically but had been occupied for as much as a million years by a distinct lineage of Erect humans. The superabundance of wood and bamboos in the region has led some archaeologists to suggest that plant material dominated their technology. Plants are also likely to have figured very prominently in their diet. Large blocks of mature forest would have been uninhabitable for humans, although narrow zones close to margins or open areas could have been productive foraging areas because there are so many fruiting and herbaceous plants, and this zone is also the preferred habitat of many small and medium-sized species of animals.

The most favourable habitat for Moderns and perhaps for Erects too would have been river courses and shorelines. These were multiplied in this region by a high rainfall and a fretwork of promontories and islands. Their dependence on fresh water had ensured that all early human activity centred on streams, springs and lakes. To this extent riverside life in Southeast Asia could have resembled the moister parts of India. Shore dwelling, however, was decisively different. Marginal groups of the great continental populations are known to have foraged along seashores in Africa, on the Mediterranean and in India. In island Southeast Asia the resources of the hinterland were less accessible and reliable and the environment itself more forbidding than the thousands of kilometres of shoreline snaking in and out of the bays and headlands.

As in Africa and India, Modern humans displaced their predecessors, inherited their land, their food supply and even, perhaps, part of their technology. Just as important, they probably acquired with great rapidity a large part of the ecological knowledge and cultural adaptation that had been so long in the making. It is inconceivable that Erect humans could have occupied island Southeast Asia for a million or more years without exploiting any of the food resources of the foreshore and it is possible that at least some of the shell middens from this region were made by Erects.

The imitation of activities such as shell gathering is hardly of great significance, since even storks, thrushes and mice make small shell middens. Nonetheless, the cumulative benefits of learning many small details of hunting, fishing and food-gathering techniques would have been a short cut to local adaptation. This development could have been greatly helped and speeded up by the direct imitation and adoption of a pre-existent and well-tested culture but there is in any case a natural progression from harvesting tropical rivers to harvesting tropical seas. The rise and fall of rivers offers seasonal gluts of fish, terrapins and other reptiles, freshwater mussels, crabs, waterfowl and their eggs. Most of these prey species have marine equivalents and can be trapped or speared in the shallows, especially around rivermouths, in mangrove swamps or on the foreshore. The single most important food here would have been molluscs. Easily collected even by the burdened, the very young or the elderly, shellfish occur on tropical estuaries and beaches at such high densities that they have been estimated to be the most concentrated source of residual protein to be found in any habitat, anywhere.

In a 1960s study of molluscs and their Aboriginal predators (principally women and children) Betty Meehan found that shellfish were harvested throughout the year and that, although many species were collected, one clam species, *Tapes hiantina*, made up most of the catch. On any gathering day one woman collected an average of 11.5 kg of shells or 2.4 kg of flesh, but when pressed they could collect 43 kg each, yielding 9 kg of flesh, enough protein and energy to feed three or four people. Meehan and Stephen Davis, another Australian anthropologist, have provided complete and detailed records of year-round seashore subsistence in Arnhemland. An abbreviated calendar based on their records is shown overleaf.

At a localised individual level the incentive to foray further into the sea may well have been fuelled by periodic bonanzas. Droughts, neap tides and other very low tides could have enabled waders to discover nesting sea-birds' colonies or turtle hatcheries that were normally out of reach. Likewise, concentrations of fish, dugong, octopus and other prey on outer reefs and shallows could have stimulated ever more adventurous essays in harvesting further out to sea.

Once rafts and boats had allowed beach people to reach offshore islands a sudden and vast expansion of range and diet became possible. Colonisation of the Southeast Asian archipelago or Sundaland might have begun as early as 120,000 years ago, but the process of expansion must have emanated, probably rather unevenly, from a very large core population in mainland Sunda. These people had thrived on a winning combination of shore and marine resources. Localised or individual

Calendar of major or preferred foods in northeast *Arnhem Land (from Stephen Davis,* Man of all Seasons, *Angus & Robertson, Sydney, 1989).*

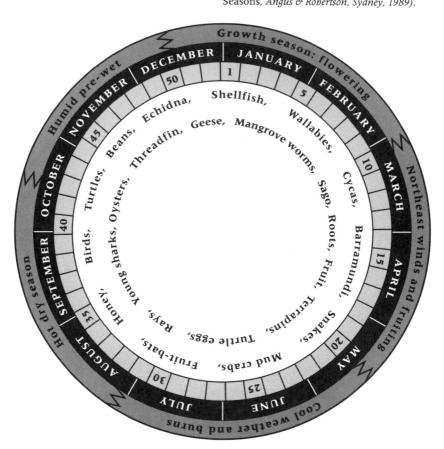

incentives are less relevant to this expansion than the internal dynamics of a well-fed and rapidly growing population that was also subject to the usual ups and downs of climate.

Every major alteration in global climate has incalculable biological effects. Thus the settlement of Australasia would have been made somewhat easier when sea levels were lower and shortened the distance between islands so that humans could island-hop until they reached the joined Australian-Papuan landmass or Sahul. Sandra Bowdler, an Australian prehistorian, considers that the colonists of the Sahul landmass initially maintained their littoral economy. As populations extended they would have followed the coasts at first, only venturing into the hinterland up rivers. Here their established water-based economy was still operable and allowed the first accommodations to be made to inland ecology. Only when population pressure increased and the climate

136

encouraged it would the first Australians have learnt to live in the outback.

The potential prey at that time included numerous species of large kangaroos and a rhino-sized wombat among some forty species of large marsupials, giant emus and several colossal reptiles. Aboriginal expansion in Australia is known to have coincided with an increase in the incidence of fire and this may have played a big part in opening up difficult woodland to make seeing, walking and hunting easier.

In the tropical north of Sahul, rising sea levels isolated the island of New Guinea from Australia about 60,000 years ago (there have been several reconnections since). The character of human subsistence and evolution was rather different in the two major landmasses of Sahul. In Australia, very thinly spread coastal pioneers fathered an entire continental population, whereas the countless seashore settlements of island Indonesia and Melanesia would have added up to a much larger population with an essentially linear and interrupted distribution.

In spite of its scatter, this Indonesian and Melanesian diaspora shared a shore-based marine economy. This economy would have caused a rapid increase in population and a huge expansion in geographic range. The first was fuelled by tapping into virgin resources, the second was helped by the mobility that came with the use of watercraft.

By an accident of geology there is a reliable record of the Melanesians' first arrival on an island and of some significant ecological changes induced by their predation. At Matenkupkum, in the island of New Ireland, middens preserve a layered record of shellfish foods from the time of first arrival (about 32,000 BP to some 21,000 BP). Geological uplift has countered rising sea levels and saved this and some other sites from being drowned.

To begin with people harvested the largest individuals of the largest mollusc species. By contrast, the more recent layers contain few large individuals, and large-bodied species make up a smaller proportion of the midden. This might be thought to represent the attrition of a resource, but ecological studies on molluscs suggest that predation has the effect of increasing the diversity of species. The biomass of molluscs at Matenkupkum might not have changed very much but one benefit from a larger spectrum of species could have been a more reliable and sustained harvest for the shell pickers.

Aboriginal boy fishing, Lloyd Bay, Cape York.

Another interesting detail of the New Ireland middens is the absence of cuscus (a leaf-eating marsupial) in the lower layers. This animal is very abundant in the forests of New Guinea and provides a vital fall-back food during months when the sea is too rough to go fishing. At the time when cuscus bones suddenly appear in later New Ireland middens, the sites show signs of being less transitory and both marine and terrestrial foods are used more intensively. Since the cuscus are most unlikely to have colonised the island unaided, they mark, perhaps, the first example of a man-made introduction. Although the evidence is sparse, it is not fanciful to envisage islanders spreading and introducing organisms such as cuscus, coconuts or yams to augment their diet on outlying islands and atolls.

Whether it began incidentally (cuscus are good 'escapers' and so are coconuts and yams in their own vegetable way) or as a deliberate planned human decision, this sort of manipulation contains the germs of agriculture and husbandry. In any event, a wide span of ecological knowledge and consistent attention to the life-history details of many small-sized but abundant organisms would have been the norm for shore-dwelling islanders. Combine this with mobility and they would have had no equal in the thoroughness and consistency with which they extracted a good living from South Pacific shores.

For a long time the communities closest to the mainland managed to combine aquatic foods with big game. There are no really early sites, but in northeast Sumatra a Hoabinian site on the Tamiang River (and about 15 km from the present coast) has remains of crabs, fish and tortoises as well as of deer, wild pigs, rhinos and elephants. Pestles and mortars had been used for processing seeds and nuts. In southwest Sulawesi (Celebes), remains of wild pigs, babirusa, anoa buffaloes, monkeys, rodents, bats, cuscus, giant tortoises and lizards augmented the great number of molluscs eaten. These islanders also gathered grass and sedge seeds and the fruit of figs and other forest trees.

The mainland of Southeast Asia and the larger islands offshore are the region in which plants first overtook animal foods in importance and at an earlier date than anywhere else. The Niah cave mentioned earlier included the remains of dominant forest mammals but it is possible that these were no longer staples. In areas of heavy rainfall vegetable foods, mostly those growing along forest edges and in openings, on riverbanks

Sago palms are felled and the pith cut out with an adze. Starch is washed out of the pith using the leaf sheaths as containers and furrows.

and shorelines, assume much greater importance, and some of these have been identified in other, much later, cave sites. They include kernels of the betel-nut palm, butternut, candlenut, pepper and almonds. Legumes include types of pea and bean and there are traces of bottle gourds and water chestnuts. Plants having no hard parts, such as yams or the pith of palms, have not been preserved but are certain to have been of great importance.

Sago palms would have been one of the core staples in Southeast Asia and a single trunk can yield 350 kg of starch. In many areas it may have been the only staple. Until very recently the Motu of southeast New Guinea mounted elaborate annual trading expeditions to the Papuan Gulf in order to obtain it. In the gulf itself, sago, sago grubs and small fish made up the largest part of the people's total diet. Although it is normally pulped, washed and dried before eating, at least one tribe used simply to chew the pith and spit out the fibre; they ate sugar cane in the same way.

Low-lying forests close to the shore are also exceptionally rich in fruiting trees which to this day attract primates, pigs, civets, bats and hornbills. Perhaps in response to the pressure and variety of dispersers, enormous quantities of fruit have been evolved, sometimes of great weight (up to 30 kg in the case of jackfruit). Durian trees, belonging to twenty-seven species, shed fruit weighing between two and four kilograms, and humans seeking a larger share of the crop still have to camp under the giant forest trees to deter other animals. Indian almonds, rambutans, lychees, Malay apples, a variety of fruiting cloves, nutmeg

140

Climbing a coconut palm.

and the citrus family all originate in this region. This widespread resource which had previously supported large populations of arboreal primates, pigs and elephants, was there for the taking from the very first intrusion of humans into the region.

Animal species eat, carry or drop fruit in different ways, so that seeds carried by pigs and monkeys as well as hornbills get more widely distributed than those solely dependent upon, say, the orang-utan. The earliest humans in Southeast Asia can be assumed to have preferred river courses and especially rivermouths (all Erect human sites tend to be from lowlands). Some prime food plants, such as the durian, are also restricted by altitude (about 700 m) and many fruiting trees grow best on levees or along coast littorals because they are 'planted' in good silt by floods.

It is obvious that aquatic environments offer rather little in the way of plants. Water lily roots, shoots and stems are choice foods, but mangrove fruits are tolerable only when there is nothing much else. By contrast, fertile, well-watered soils on overspill flood plains and seasonal swamps grow many green annuals, ephemeral herbs and bean-bearing legumes; vines, lilies, arrowroots and sedges have edible seeds, corms or stems. On the shore and in low-lying valleys there are coconuts as well as sago-producing palms, superabundant rattans with edible shoots and fruit and the ubiquitous fruit-bearing pandanus. On the river levees and banks, wild mangoes, Indian almonds, jackfruit, wild figs and bananas would have grown in great profusion. Although there would have been few natural savannas in island Southeast Asia, seasonal and local differences in climate would have ensured considerable ecological diversity. For example, plants that die back or lose their foliage over the dry season often renew growth from underground tubers or corms. Local yams, taro, arrowroot, elephant yams, bush potato, asparagus bean and types of cowpea would have had their natural role of storage organ diverted to serve as a carbohydrate store in the prehistoric human larder.

Some of these tubers or corms anticipated attack by developing toxins; in such cases their use by humans would have been dependent upon processing that involved peeling, soaking and, above all, baking. In any case, most raw roots and tubers are pretty unappetising and their collection can be correlated with the use of fire. Root-eating would also have signified a great expansion of the food supply because yams especially are superabundant throughout the monsoon tropics.

Wherever human groups congregated for any period of time they would have burned, cleared and opened up the surroundings. Since only a few roots and tubers flourish under a closed canopy and many will regenerate from severely damaged fragments (especially yams), people would have noticed yam regrowth not only in pig rootlings but also in their old middens and litter when they returned to or revisited previous camp sites.

Distinctions between individual plants that have propagated naturally and those that spring up from human or animal interference become less meaningful when it is remembered that the yam-eaters would have known their foraging area intimately. In the Andaman Islands, for example, where wild yams are the main vegetable staple, harvesting of them before they mature is prohibited by religious sanction. In such a situation people know and remember the location of their first 'crop' and dig it up only at the appropriate time of the year. The annual cycle of their prey and forage are closely observed and well known to hunters and gatherers wherever they have been studied and we can be certain that both Erects and their Modern successors were equally acute. After all, small carnivores, rodents and even birds revisit caches long after they were first made.

It can be assumed that early hominids observed ecological processes and patterns, remembered events and reserved their action until later. Thus individuals could have found their way back to already known plants in order to harvest them and companions might very well have gone along too. Where the Andaman Islanders differ from this simple delayed response is in their manipulation of group behaviour so as to maximise a predictable recurrent benefit. For this, both complex speech and the socially manipulative mechanisms of beliefs and sanctions are necessary. Here the Andamanese exhibit typically Modern human behaviour.

Food plants that regenerate in middens or find fertile and sunny seedbeds on the margins of camp-site clearings or in latrines expose the details of their life cycles to human inspection; they invite an awareness of dependency. They are not *encountered*, they are *known* in familiar detail and it is possible that where camp sites were regarded as being 'owned' by a particular group, the edible plant growth around it was also regarded as being owned.

Many nomadic people regard waterholes and camp-sites as property and the spacing out of nomadic bands would also have depended upon exclusive or priority use of key resources in the heart of their annual home range. These resources could have included many more plants than a few home-grown yams. For observant humans (and hunter-foragers had to be observant) the rapid growth of yam vines from the leftovers of their tubers would have been a near-instant demonstration of the beneficial consequences of messy dining.

Now view the situation from the plants' point of view. For them usurpation of the role of top seed disperser by humans generally narrowed the range of potential beds their seeds might fall in. Those plant species that could still propagate in spite of (or because of) peculiar human behaviour would benefit the most. Among the assets would have been fewer seed predators around an encampment.

It has been suggested that 'mainlanders' in Southeast Asia were more involved in opening up forest than their island-dwelling neighbours. The clearing of significant areas implies high concentrations of people and this in turn requires exceptionally rich food resources. When residence in a small area is fairly continuous, the use of wood for cooking fires alone eventually begins to consume an ever-widening circle of the woody vegetation. Among the profusion of plants that spring up in clearings are many that are useful, but which ones prospered must have been more than a matter of chance. Prehistoric people would have been well aware of competition between plants, and can be surmised to have intervened on behalf of useful plants when these were being outcompeted by 'weeds'. Indeed weeding may have been one of the early steps on the path to true horticulture.

One implication of this is that forest clearings were altered to favour the many herbs that spring up in each clearing. Prominent among these were leguminous plants such as peas and beans, bottle gourds and yams. This proto-horticulture might have lasted for tens of thousands of years before becoming formal swidden agriculture. Gourds, like ostrich eggs and animal skins, were probably used as water-containers and carriers from very early times and many natural gourds have hard, durable shells which can be used with a small neck opening or bisected to form bowls.

Many food plants have been subject to intense competition. Wherever feeding grounds were especially choice and concentrated there would have been strong incentives to catch other competing animals. One effective method would have been to corall the area, interrupting the brushwood fence with traps and snares on every incoming path. In this way one resource was better protected and another gained. It is possible that some of the earliest 'fields' and 'farms' began in this way.

Many of the wild plants that were still collected by hunter-gatherers in Australia until very recently had long been domesticated or semidomesticated in Southeast Asia – yams, arrowroot, palms, including pandanus, cycads and various nuts including *Terminalia* and *Eugenia*. In Arnhemland the parsnip yam was harvested carefully, leaving portions of rootlet attached to the tuber top, so that the yam would regenerate the following year. Yams in this region used to be deliberately planted on offshore islands to ensure future supplies.

The process of domestication may sometimes have had more to do with an increasing density of people than with any change in the character of the plant. In Cape York, marks used to be left so that wild yam vines could be recognised as property and wild honey hives are marked by many cultures to reserve them for their mature stage. Since the resources near the centre of a group's territory would be acknowledged as closed to outsiders, it would have been towards the outer margins that marks would have helped assert and define boundaries.

A picture of how the land and territory were perceived by a foraging community emerges from Walter Roth's description of Queensland about a century ago.

Independently of the greater or less extent of the country over which the community as a whole has the right to hunt and roam there are definite territorial divisions, certain tracts of country for each family, each such tract bearing a distinct name. In the same way as an European knows what vegetables, shrubs or flowers are growing in his garden, so do the natives have a very fair idea of the amount and whereabouts of any special growth of edible roots, fruits and seeds, as well as the particular haunts of the various animals and birds found on their own particular piece of ground. For one family or individual to obtain vegetable, fowl or meat without permission upon the land belonging to another family constitutes trespass and merits punishment. This, however, is not of a very serious character, unless a non-tribesman is concerned: a slanging match, with both parties indulging in obscene epithets or a spearing in the leg. Trespass is, after all, but rarely committed, considering that, on account of their very hospitality when one family experiences a superabundance of food of any description, its friends and neighbours are generally invited to come and partake of it. For a non-tribesman to trespass means death.

Even if a speared leg is considered 'not very serious', there were obviously few incentives to trespass, and the popular image of foragers wandering at will in search of what they might happen upon was never a realistic picture of a prehistoric forager's existence. In many countries plants are deliberately signposted or artificially propagated as boundary markers and these may be economically useful or simply conspicuous.

Agta child fetching firewood (Photo J. Kamminga).

Such practices have prehistoric beginnings and group-territory definition could be an important dimension of plant domestication. Wherever
human numbers began to exceed natural carrying capacities those living
in areas with an assured surplus would be particularly concerned to mark
out and assert their rights in the face of mounting competition.

In similar fashion prehistoric people could have influenced both the

Prehistoric humans gathered and pursued their activities in a variety
of well-defined sites, often centred upon a single food source. Camping
in fruiting durian groves is the contemporary continuation of a practice
that must go back to early hominids. Eating and excreting seeds in the
vicinity have favoured durian propagation and helped to perpetuate the
regrowth of trees in the same locality over immense periods of time.
Where parent trees grew in repellent or dangerous places their fruits
might have been collected or transported to processing sites that were
more comfortable for the collectors; while this may have been effective
in dispersing seeds more widely, differences in the chemistry or soil
fertility at a processing site could have inhibited propagation of all but the
rarest of seeds. If a seedling survived that was genetically different from
its parent stock, the hominid that transported it was unwittingly its
'breeder'.

In similar fashion prehistoric people could have influenced both the

genetic strains and the growing sites of many food plants. Since a convergence between the plants' needs and the dispersers' convenience was implicit in the beginning of the association, it is reasonable to suppose that soon after they first arrived in Southeast Asia Erect humans began to influence the ecology of their food plants. The very great age of plant–human relationships here is evident from the fact that domestic strains of banana and breadfruit have lost viable seeds in their fruit and have come to rely on vegetative reproduction from cloning or cutting.

A subsistence in which plants were a major, rather than a supplementary, proportion of the diet would have come to be the principal characteristic of mainland Modern humans soon after their entry into Southeast Asia.

Chinese historical botanist Dr Li has identified two distinct regions where cultivated plants originated in the Far East. Northern China and the Southeast Asian archipelago are and always have been totally different ecologically, but the emergence of a distinct human population in the south at a very early date would have meant that early horticulture developed out of very well-established traditions, in which the domesticates were already familiar as wild foods, their habits and seasons known in the greatest detail. The concept of possessing plants and the places they grew in, various interventions to promote growth and possibly some weeding must all have preceded horticulture. What was lacking was the idea of systematic planting and tending.

That idea must have arisen in areas where growing conditions were very favourable, where human densities were high and territories small. Also, I think it is a fair bet that horticulture was begun by women because they were more tied to the home base by their children and were better placed to tend plants while the men were away trapping, fishing or hunting. It is also possible that its spread was delayed because the neighbours of horticultural societies would have been slow to adopt or imitate practices characterised by sedentariness, monotony and a much narrower range of subsistence activities. The sustained work of digging, weeding, guarding and processing was unlikely to have been adopted by nomadic foragers except under duress.

It is for these and many other reasons that agricultural and foraging cultures tended to maintain their distinctness up to the present. Where there were racial differences, as was often the case, the economic separation was reinforced by cultural prejudice, further inhibiting the spread of gardening to non-horticulturalists. The anthropologist K. Endicott has described a Malay explanation for these people's distaste for Batek Negritos (black forest foragers) in terms of a legend that the Batek derived from a ball of dirt scraped off the skin of the first man, Nabi Adam. Similar contempt for foragers is widespread among agricultural people the world over.

The spread of horticulture is therefore likely to reflect expanding population pressure from groups that could rear large numbers of children. The adoption of horticulture by established foragers would have taken place only when other options were closed, by which time their small numbers ensured rapid genetic absorption.

While cultural and economic divergences would have been relatively rapid between gardening islanders and beachcombing shore dwellers, this need not have implied much racial difference, especially on islands. A very telling picture of divergence between people of the same stock has come from the Andaman Islands where the (now extinct) coastal Arioto tribe used to call their neighbouring inlanders the 'strangers', or Jarawa. Until very recently the Andaman Islanders preserved, in almost total isolation, a time capsule of the main modes of existence followed by the Banda strandlopers, one on the beaches and in the sea, the other along the forested streams and valleys inland. In the Andamans, the boundary between them was maintained by mutually incomprehensible languages and by a fierce antagonism in spite of their close and continuous proximity. The Italian anthropologist L. Cipriani, who lived with the Andamanese between 1952 and 1954, wrote

> They represent today the survival, almost unchanged, of a culture once to be found all over Oceania, spread thousands of years ago by canoes hardly different from those the Onges use today. They carried the unknown seamen from island to island, bearing with them a way of life. . .

The consistently rich resources of the islands and the strongly territorial behaviour of their inhabitants would have combined to reinforce isolation and conservatism. Thus the outside world's first contact with the Andaman Islanders only began in 1790. (A tradition of repelling all invaders probably explains their sustained isolation in such a busy sea as the Bay of Bengal.) The original islanders were subdivided into five tribes of which three were coastal and two inlanders.

Great Andaman
  Arioto – coastal littoral only (extinct)
  Eremtaga – inland (extinct)
Middle and South Andaman
  Jarawa – inland (nearly extinct)
South Andaman and Little Andaman
  Onge – seafaring and whole of Little Andaman (extant)
North Sentinel Island
  Sentinelese (nearly extinct)

The islands have a very wet monsoon climate (3,750 mm of rain a year) and were originally covered in dense forest before clear-felling for timber began. Recent surveys of the flora have recorded a rich variety of over 7,000 plant species. The main surviving group, the Onge, combined seashore and inland foraging (but no point on Little Andaman is more than eight kilometres from the sea). Most land foods were collected by Onge women and children using baskets and digging sticks. The wild fruit that they gathered included bananas, jackfruit and pandanus dates. Stems, green vegetables and several types of roots and tubers were also eaten. The most important plant food was tubers of the local wild yam *Dioscorea glabra*, but these were seasonal and not used as stores except during the brief period of movement from one hunting ground to another. Very large shell middens show that molluscs have been a consistently important food for many thousands of years. In early times these were baked but boiling came in with earthenware vessels, which were introduced from Burma perhaps two or three thousand years ago.

When the Onge were foraging inland, a small locality was scoured for foods and a temporary shelter (called korade) was built. Once the area was exhausted of resources the group moved on to another part of their territory and put up other korade shelters. Well-made communal stilt houses called bera were erected on ridges or higher ground and endured much longer. The Bera used to serve as a central focus for group activities.

Honey was of great importance to the Onge and they are said to have followed the patrols of honey-guide birds (*Indicator*) to discover hives. The most favoured type of yellow honey was harvested in great quan-

Andaman islander fishing with long bow.

tities during the dry season, while the brown honey of another bee species was gathered during the rains. While the men gathered yellow honey, the women and children took advantage of the low tides and calm seas to beachcomb for small fish or other reef fauna.

Onge men, when not collecting honey or hunting in the forest with spears and bows and arrows, specialised in seasonal fishing. The very numerous wild pigs used to be hunted without dogs (which were unknown before 1857) by small parties of males or by a single skilled and experienced hunter. The remains of pigs appear in middens at the same levels as the first pottery and probably have a similar date of introduction.

Onge women fished the abundant forest streams, river mouths and beaches for crayfish, crabs, molluscs and fish, using small chika nets (strung out on a circular twig frame) or sometimes shooting fish with bows and arrows. The men too used bows and arrows but also spears to impale sea fish. This could be done in the shallows or from canoes. For dugongs and turtles a harpoon was used with a detachable head (tied by a cord to the shaft, which also served as a float and marker buoy).

Turtles were hunted from canoes and harpooned while they slept on the surface, while mating or when corralled within the reef during storms. However, the harpoonists preferred still evenings and nights, using torches made of resin-soaked tightly bound palm leaves. The torch was slung beneath the harpoonist's platform on the prow, while the steersman sat in the stern with a paddle (or in shallower water stood and poled). Occasionally a third man might squeeze in amidships. Beaches were avoided at night for fear of spirits but were combed avidly during the day for signs of turtle nests. Turtle eggs were baked or boiled and the turtles too were either roasted alive over a slow fire or boiled piecemeal.

Fires were formerly tended very carefully and carried on any trip, however short and temporary, as well as on long journeys that involved a change of place. The anthropologist Radcliffe-Brown described the Andamanese as being skilled in selecting and using types of wood that would smoulder for long periods without either going out or flaring up. This careful conservation of live embers was necessary because they had lost (or never had) the knowledge of making fire on their own. By the turn of this century, however, matches and fire sticks had become familiar. Cooking was done in pits with a fire placed above rather than below the food.

When first discovered the Andamanese allegedly had no knowledge of coconuts. If that is true, the worldwide distribution of coconuts on tropical shores may have been a relatively late human-assisted dispersal. It is often assumed that coconuts owe their wide range to the ability of their nuts to float in the sea. Instead, they might originate from a fairly restricted area. Since there are no wild coconuts left, their area of origin must be conjecture, but the distribution of insects dependent on coconuts and coconut crabs points very strongly to a South Pacific origin. If coconuts were deliberately propagated and carried by pelagic humans that event could have followed after the settlement of the Andamans and also postdate the islanders' opportunistic mode of hunting and food gathering (unless an epidemic eliminated the palms). This could have extensive implications.

If the Andamanese mode of subsistence once typified most littoral dwellers throughout the Indonesian arc to the furthest reaches of Melanesia, it contrasts strongly with most of today's inhabitants in this vast region, who are now highly skilled gardeners and fisherfolk. Even though the origins of their agricultural practices are likely to be the earliest on earth, many of their contemporary crops are known to be of relatively recent origin and coconuts are likely to be among them.

Using a relatively rudimentary technology, the Andamanese could harvest quite adequate resources. Even in the 1960s, long after their forests had been degraded by clear-fell logging and their society marginalised, disease-infected and impoverished by an ever-growing immigrant population, they were able to keep up a good diet without external supplements. A nutritional survey made by Dr Bose in 1964 sampled their diet at a time when some of their prey, notably dugong and turtles, were known to be scarcer than in the past. The survey was made for one month during their dry season (December–January) when normally common foods such as fruit, molluscs and turtles were scarce or unseasonal. Despite all these limitations, a community of forty-one individuals consumed some 1,100 kg of pigs, fish, crabs, wild vegetables and honey. The harvest represented about one kilo per day for every man, woman and child (and about 1,760 calories of energy intake).

The self-sufficiency of their economy makes it unlikely that the Andamanese had lost skills through forms of cultural stranding or trauma. The absence of fire-making, of sailing boats, of man-made salt, of any form of cultivation, and the uniqueness of their language, all point to the maintenance of an ancient way of life by a relict population.

If their parent culture were the first to develop a means of colonising and exploiting tropical islands and coasts, this history could give the Andamanese a special significance. Instead of calling the Andamanese primitive, imagine their culture as a novel and unusual fashion of exploiting the environment. Visualise the huge food supplies made accessible by boats, spears, traps and nets. Imagine the rapidity with which such a well-fed and mobile people could multiply and spread along empty beaches and down uninhabited coastlines. Consider what confidence these seagoing fisherfolk would have brought to their possession of a well-tested technology, the best of its time. Picture them, too, skirting the subcontinent of India. The prehistoric ancestors of the Andamanese must long ago have been as thoroughly subsumed in India as the living islanders are now being integrated into today's Indian mainstream (the process may not be complete but it has gone a long way in not much more than a century).

Beyond India, a similar subordination to mainland people might have taken thousands of years to occur, but could such a seashore economy have extended that far anyway? The marine food resources of south Arabia would scarcely have differed from those along the Indian and African coasts. At times, some stretches of the littoral might have been short on fresh water, and edible plants might have been rare along much of the Omani coast but it would be false to invoke a shortage of food as the barrier to westward expansion by a pelagic culture.

In Africa, the Australian sequence of unimpeded expansion into an empty continent was not possible, because the mainland was already occupied by a population of big-game hunters. The large-animal hunting economy that sustained the earliest human societies still supported small groups of people in Africa until this century. It has lasted longest on the drier southern and eastern sides of the continent where it was overlaid, one or two thousand years ago, by pastoral systems that largely replaced wild herbivores with domesticated ones. Today's pastoralists and the vestigial hunters of southern and eastern Africa are, for the most part, quite distinct from the majority of peoples of central and western Africa, in spite of extensive mixing and interpenetration over tens of thousands of years. These pastoral and hunting people number some millions and are highly diversified but they can still be grouped in the two broad categories of Khoisan in the south and the variously named Hamites or Cushites of north and eastern Africa. The sequence of events that is suggested here has been constructed from meagre clues and put into a time frame where the dates are admittedly uncertain but at least some of the ancestors of the two regional groupings, northern Cushites and southern Khoisan, may have been the inhabitants of eastern Africa's interior at the time when Banda strandlopers first colonised the littoral. The equator would have roughly demarcated the boundary between them.

For the new coastal colonists the most favourable coasts lay below the equator. The absence of permanent fresh water would have been a major constraint on settlement and it is only from the equatorial Juba River southwards that localised rains maintain a narrow and discontinuous beading of forest patches all the way to the Limpopo River. The eastern coast of Africa would have been as rich in marine resources as the shores of southern Asia. Much marine life spans the entire Indian Ocean, so that the sea harvest and techniques needed to gather it would scarcely differ from one continental edge to the other.

It is probable that a few aboriginal African people exploited the shoreline but the new arrivals would have been overwhelmingly superior to these small peripheral groups in their capacity to extract a rich living from the seashore. In all the tropical and subtropical habitats, the littoral would soon have become the exclusive territory of immigrant Banda.

Beyond the Limpopo, however, the situation would have been rather different. Here a relatively large and well-established population had led a coastal existence for tens of thousands of years. They seem to have done without boats and fishing but the remains of molluscs, seals and sea-birds, augmented by small game, turn up in a long succession of middens and cave deposits all along the Cape coasts. The tropical-adapted Banda might not have ventured very far south of the Limpopo River, because this temperate region was the territory of a relatively large population of well-adapted competitors.

The tropical east coast forests do not stretch inland very far. Forests recur on hills and mountains and line every river and stream that empties into the sea, but behind and beyond is a mosaic of very dry steppes, thickets, woodland and savanna. Then as now, plant foods could have been scarce in the dry season and the abundant large mammals would have been the main prey of the continental people inland. Here, the aboriginal Africans maintained their sway over the whole southeast-ern interior of Africa. This was a region that was sensitive to quite small shifts in climate. During dry phases an arid or semi-arid corridor would have joined Somalia with the Kalahari, but always there would have been broad bands of prime big-game country on either side. This is where the great world heritage and national parks lie – Tsavo, Serengeti, Ruaha, Mahali, Luangwa and Etosha. These modern vestiges of the great Pleistocene mammal communities offer a glimpse of both the abundance and the variety of large mammals. They also illustrate ways in which all the lesser life, animal, vegetable and human, is subordinate to the ecological impact of big beasts. It is scarcely surprising that the rise of humans in Africa was fuelled by the use of a killing and trapping technology that turned them into superpredators at the top of the food chain. Such a long and effective tradition of living off these vast resources was not easily abandoned.

It was into this primary mode of human existence that the Banda strandlopers intruded when they began to expand into the interior.

The resources of the African coast are finite. Once all the habitable foreshore had become settled in its entirety, the natural course of population expansion would have been up the larger rivers. Here, fish and reptiles would have been the main aquatic staples, while water-edge animals and plants provided the rest of the diet. The primary orientation of the fisherfolk was to *water* animals and *land* plants. On the coast, plants were likely to have played a subsidiary role. Up river, where the aquatic habitat narrowed and its faunal diversity decreased, the emphasis would have shifted so that plants, especially those growing close to the riverbanks, would have become a more prominent part of the diet.

The Banda had developed an economy that used a wide range of small items intensively. Where mega-hunters could only exist at moderate densities in the best game country (because they and their prey were nomadic), the Banda would have sustained larger communities at appreciably higher densities with much less movement. If one economy could support larger numbers of people on the same piece of real estate, this would have ensured that simple demography eventually settled the outcome of the competition for food, water and space.

If the Banda became established between 75,000 and 55,000 BP much of this was a period of ecological instability. Well-watered habitats may have fluctuated but, in spite of this, may have been less perturbed than dry savannas where large mammal populations were more prone to cycles of boom and bust. This instability could have had knock-on effects for the mega-hunters and tipped the demographic balance still further in favour of the Banda economy.

Such a historic incursion should have left some marks and there are indeed archaeological traces suggesting that a more plant-oriented culture once developed in the region lying between the Zambezi and Limpopo rivers. This is the Charaman industry described earlier and the Charaman combination of heavy and crudely made tools and smaller, more refined blades is consistent with the exploitation of at least two very different classes of food.

At least one authority has identified the function of the heavier tools as hoes or picks and it has been pointed out that their distribution coincides in equatorial western Africa with the 'yam zone'. Yams have proved to be the most reliable source of bulk food in many tropical regions with a rainfall of over about 1,100 mm, and they are very important for all living hunter-gatherers that inhabit equivalent yam zones in Asia, Australasia and Africa.

Unfortunately, no trace of a yam has ever been preserved in an archaeological site, so their ecological dominance and botanical bio-geography must be the main guide to their role in prehistory. Wild yams are very common all down the eastern African coast where a nontoxic species, *Dioscorea hirtiflora*, is still gathered and eaten. This yam extends inland and is a common dry-season bush food in Mozambique and Malawi, where a well-marked division of labour (said to be age-old) governs its harvest.

Women and children are the principal foragers for yams (although the men may sometimes have located them). This specialisation is extremely widespread on all continents and continues to apply also to people who now forage for wild foods only on an occasional basis. It seems to be part of a broader division whereby women ensure that perennial reliable staples are always available for the survival of their children. In hunter-gatherer societies women have focused on items that could be collected in spite of the impediment of babies and young children and therefore tend to be safe, sessile types of food. Any prey that was 'not for children' was the males' prerogative. Large, active, fast, difficult or dangerous prey was, by its very nature, less reliable to harvest.

It is possible that the efficiency of large game-hunting techniques combined with prey abundance had conspired to make established mega-hunter societies more specialised than the broad spectrum foraging of strandlopers. Molluscs and turtle eggs on the shore are the equivalent of yams inshore. Both are traditionally women's work and require long, patient digging, but both ensure large quantities of food.

Yams average 5–10 kg in weight, but some varieties (only a minority of the 600-odd species are edible) can weigh up to 50 kg. Some are seasonal, others can be harvested throughout the year. Some species can be excavated quite easily but some of the larger and more highly prized species have long tubers that go down as much as two metres into hard soil. Excavating these yams would be best achieved by a combination of stone picks, digging sticks and hands raking out the loosened soil. Children can be less of an impediment than a practical help in this long, slow excavating job. While the excavation and eating of yams possibly go back to the earliest fire-users (because they must be cooked), this tuber more than any other is likely to have become the staple for a distinctively new and eventually very successful lineage in Africa.

Yams also have many useful by-products, which are known by contemporary, or have been used by, recent people. In southern Africa, two wild species are used to prepare a bait which immobilises monkeys, while yam extracts can also be ingredients of arrow poison. In Southeast

Asia, both birds and fish used to be stupefied with yam poisons and a yam shampoo was used to get rid of head lice. Learning that an underground root has all these toxic properties implies a very great and long-standing intimacy with the plant.

Charaman stone picks are so crude and rudimentary that they are not easily separable from similar tools made by early hominids, but the resemblance takes little account of their likely context and function. Already burdened with children and some form of carrier bag or sling, women would not normally lug around heavy stones as well. It is more likely that every time a good yam patch was found, a suitable stone was fashioned sufficient to the task in hand and then discarded. That this was probable can be deduced from the picks' tendency to belong to local parent rocks and not to some distant or unusual mineral. The improvisation of a throw-away tool could explain why an apparently primitive tool type could coexist with much more finely worked tools used for other purposes.

An unusual site at Kalambo Falls on the Tanzania–Zambia border consists of waterlogged soils that have preserved edible fruits, a wooden club and digging sticks, burnt logs, a fireplace and an arc of stones that may mark out the base of a shelter. Stone artefacts at Kalambo belong to two industries, late Acheulean below and very different tools above. These include crude picks as well as smaller, finer tools. Fruits preserved at Kalambo include those of the palmyra (*Borassus*), a palm that is claimed to have 801 uses in south India, wild persimmon (*Diospyros*), sand apple (*Parinari*) and Guinea pepper fruit (*Xylopia*).

Charaman-like cultures seem to have spread into an ecological and geographical area that showed virtually no signs of previous human occupation, the equatorial forest zone. The period between 60,000 and 30,000 BP saw many climatic oscillations, the drier phases of which would have degraded parts of today's forest belt into a mosaic of gallery forests and savannas. Whereas the initial adaptation of African Banda may have been to riverine forest dwelling, the shifts in climate and increasing population could have pushed some communities towards the savannas but induced a decisively forest-adapted life in others. In the region of first contact there would have been a prolonged interaction between two peoples, two cultures and two environments.

This early interaction may have left both cultural and genetic marks on subsequent inhabitants of the area. The Pomongwe and Bambata caves in the Matopo Hills in Zimbabwe have quartz blades, burin points and scrapers associated with small animal prey such as hyraxes, duikers and other small antelopes. The sites have been dated at about 42,000 years BP and this 'Bambata industry' has been named after one of the caves. Unlike the South African industries of this period, pestles, grinders and pounding stones appear. These imply that seeds and nuts had become a more systematic large component of the diet – possibly an effect of the interaction between immigrants and residents. Nonetheless, the broad character of this culture remains that of an open savanna hunting society.

Further to the north, and within what is now the forest belt, the 'Lupemban industry' became very widely distributed at much the same period. Tools for processing plant materials are common and include bored stone balls that are a type used to this day to give extra weight to wooden digging sticks. A broken digging-stick stone has been unearthed at Chabuage in northern Angola and dated to about 40,000 years.

There are numerous tubers, corms and roots other than yams to attract the diggers. Central Africa is particularly rich in edible root crops and several would have featured in prehistoric diets for no less reason than that they are common and their edibility demonstrated by the diggings of other animals, such as warthogs, bushpigs and blesmols. These would include *Fockea*, *Raphronacme*, Tsin beans (*Coccinia*) and cowpeas (*Vigna*) (all of which are prominent in the diet of the modern Khoisan). The tubers of asparagus beans and their young pods are extremely rich in protein, as are peanuts of the Bambara earthnut. Other common legumes that provide nutritious pulses are several species of wild African pigeonpeas. By far the most important foods endemic to Central Africa, and probably first collected and eaten in a big way by the Bambatans and Lupembans, are the millet and sorghum cereals. Two species of sorghum, pearl or bulrush millet and a large seeded marsh grass (*Echinochloa*) originate in Central Africa.

In the forest, roots would have been augmented by great numbers of nuts and fleshy fruits. Once again, forest animals would have been a guide to both the edibility and the source of forest foods. Many palm species bear edible fruits, shoots or piths, but the oil palm outdoes most in the sheer weight of dates it produces. The majority of these are consumed by birds, mammals and insects but, wherever Lupambans lived close to the water's edge, they would have been likely to have derived a high proportion of their diet from these superabundant dates. It would seem that African forest trees have competed with one another to produce the largest number of fruits; there are more than twenty very tasty and nutritious wild fruits that are still greatly sought after by rainforest people, including the *Cola* fruits (of Coca-Cola fame), the butternut *Coula,* and the four- or five-kilo pods of *Treculia*, which is relished to this day as *agusi* porridge. A very widely eaten spinach known from Australasia and Indonesia to Africa is the gymnosperm *Gnetum.* Eventually, some of these plants became wholly or semi-domesticated, but the large plant component of the equatorial Africans' diet probably contrasted with the predominantly animal foods of their more conservative neighbours in the savannas to the south and north.

One feature that served to demarcate a boundary between the people of sub-Saharan Africa and those to the north was the nature of the Sahara's fauna and flora. During dry, hot periods the waterless Sahara would have supported few people but the North African contemporaries of the Lupambans, with an industry called Aterian, lived over a large part of the Sahara during relatively cool periods. At such times, fauna and flora typical of the Mediterranean migrated south into the Sahara.

The Aterians hunted deer, oxen, Barbary sheep, goats and asses, which are not typically African, but there were also large numbers of ostriches, zebras, rhinos, elephants, giraffes, buffaloes and various antelopes, including gazelles. Wide, shallow lakes in the Sahara, full of fish, crocodiles and hippos, were fished and hunted by the Aterians but when it became warmer the Sahara dried out and the Mediterranean fauna and flora retreated back towards their northern home.

Before 40,000 BP human populations that had lived in the eastern Mediterranean began to move northwards. The ecology of these invading hunters in Europe is one of the best documented in prehistory. The European economies continued their ancient preoccupation with big-game hunting, an adaptation that was as well suited to the long dormancy of plant growth in winter as that of its predecessors in North Africa had been to the aridity of a long dry season.

This review of food and food-getting has portrayed an animal gradually expanding and increasing the variety of its foods. For the people

Map of the Sahara region during maximum pluvial conditions in Pleistocene.

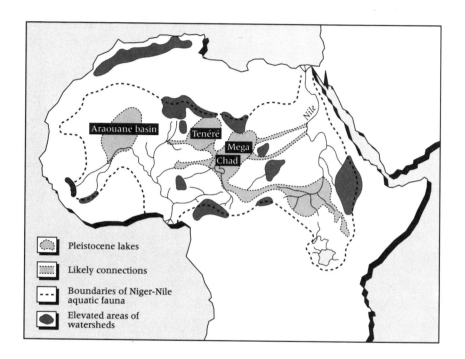

concerned, hams, clams, yams and other staples were more than mere foods – they were the Staff of Life, Gifts of the Ancestors and a central focus for law and religion. Cosmic and climatic events were domesticated by being linked with the familiar intimacies of food and food collection, and a succession of harvests marked out the annual calendar. The particular characters of regional cultures were shaped by the way they accommodated to and anticipated the workings of their local ecosystems. Understanding all these aspects of the annual food cycle would have been the main content of a prehistoric education and governed the socialisation of individuals into the group.

Modern foragers provide a guide to the quality and substance of knowledge necessary for subsistence. For example, a group of Australian Aborigines ranging over a territory of many hundreds of square kilometres of arid desert in western Australia carried a mental map of 300 potential water sources. The sites included every category of waterhole and a variety of rock clefts and cisterns, hollow trees and dew traps, all of which were known at first hand. As emergency knowledge they remembered another hundred sites from other people's descriptions. Water is just one of the many resources that were remembered in this sort of detail.

There used to be a remarkable Aboriginal artefact, the woomera, or spear thrower, that was sometimes carved or painted with a 'map' of springs and waterholes. The owner of one example named sixteen springs marked out on his woomera, which depicted them among the coiled tracks of an ancestral totemic snake of the Dreamtime. This trail meandered over sandhills and rocky ranges along the 250 km course that was real enough to be followable on an ordnance survey map. When the owner set out along the snake track he trusted to his own capacity to find his way around and feed himself, both skills learnt from early childhood. More important, he had learnt from those first forays with his father, uncle or family party that he was accompanied by other invisible forefathers. Thus he could walk alone and naked, armed only with his spear and woomera; because his family was embedded in the landscape he had inherited, here he was secure and among his own.

Theodore Strehlow, a student of Aboriginal religion, described a relationship between land and people that might have had many parallels in prehistory and very far from Australia.

> Creeks and springs and waterholes are to him . . . the handiwork of ancestors from whom he himself was descended. He sees recorded in the surrounding landscape the ancient story of the beings that for a brief space may take on human shape once more; beings many of whom he has known in his own experience as his fathers and grandfathers and brothers and as his mothers and sisters. The whole countryside is his living, age-old family tree.

I have travelled with African hunters over a territory where every single landmark, valley and hill, sometimes individual rocks, had a story attached to them, more often than not one that was intimately linked with an ancestor. I remember visiting a waterhole where the only drinkers now were elephants and antelopes. 'This was where my great-grandmother drew water every evening, and it really flowed then, not the trickle you see today.' An aged fig with elephant-wounded roots winding into the seepage was the landmark-cum-guardian of the site.

In many instances the territory or living space used to be defined in terms of water. For prehistoric and recent foragers alike an awareness of water as the most sustaining of all commodities was linked directly with those from whom all authority and knowledge flowed and all security derived – the ancestors.

To this day foragers and nomads who are very flexible about territory can be fiercely protective of springs they have inherited from their forefathers. Those sources of water that had been proved by handed-down memory to be reliable and perennial would have been identified in prehistoric experience with the ultimate security (after all, a human can survive for a month on little more than water). Everlasting water also manifests continuity and predictability in an otherwise unpredictable universe. No wonder then that springs had a sacred aura.

The observation that springs were recharged by rainfall, even by quite distant mountain rain-clouds, would have been understood by many prehistoric people. It is also likely that, because of the mode in which knowledge was transmitted, a large part of what was learnt would have had to pass through a filter of familial associations. Thus there would have been enormous local variation and elaboration of the modes in which knowledge was put across. There is some reflection of this today in a wide range of creation myths, folklore and vestiges of past pantheism (although most belong to later cultures).

Because information, beliefs and skills were handed down by means of one-to-one word of mouth and by demonstration, and could refer only to small, well-known localities, most of the accumulated ecological knowledge was non-transferable; it was expressed in a highly localised and 'ancestralised' format. Even if their knowledge of a strictly biological calendar found expression in esoteric symbolism, all foraging peoples followed life plans that integrated society into an ecological matrix.

Inasmuch as 'planning' could be separated from an existence that was wholly subordinate to natural rhythms and events, the fireside discussion of what would be done 'tomorrow' or 'soon' would have served to reinforce inherited knowledge. It may also have revealed who had the most acutely observant eyes, the most deductive minds and the best memories. During times of abundance when there was a wider choice of things to harvest, tradition would have been invoked to ease decision-making and direct people's energies. 'Planning' was essentially a discussion of tradition, in which those best able to read auspicious signs were the most likely to shape group-decisions.

Children would grow up learning to identify and locate all the us·ful plants within their home range and the habits of the animals they encountered, as they followed and helped their elders. They would have been taught to notice the stages through which living things pass so that they could anticipate when and where there would be food to gather in the future.

For migrating swarms and hidden bonanzas that arrived without warning there were 'signs'. For instance, the Gupapuyngu of northern Australia noted happenings on land, such as gatherings of large dragon-flies, the fruit-fall of the pandanus palm or the flowering of kurrajong bushes as markers for events in the ocean. The insects signified that mullet were swimming inshore, the falling nuts heralded the arrival of turtles to lay their eggs, while the scarlet flowers reminded the fishers that small sharks could be harvested in the estuaries for a few weeks.

A calendar of natural events was also a calendar for behaviour. The Andamanese prohibition on digging up yams before they had ripened was cited earlier. Comparable roles for people's behaviour towards some of their staple foods occur in many other societies. In southeastern Nigeria, eating yams before the New Yam festival was once outlawed so effectively that the agronomist D. Coursey reported Ibo prisoners volun-tarily starving to death rather than eat yam before witnessing the festival in their home village.

The origins of such drastic inhibitions could be twofold – one to guard against premature harvest, the other a recognition that some wild yams were more toxic while growing than when mature. Whatever the functional origins, a more important social effect was to bring predict-ability to group behaviour and reinforce group solidarity.

Among the Ibo, food and fertility were both the gift of Ala, the earth mother, who was envisaged as the regulator of all natural processes. Prominent among her surrogates was Ojuku or 'Great Spirit of the Yam'. Intercessions with Ala and Ojuku to grant good harvests and children were the responsibility of the *osu*. For the Ibo, an *osu* was a sort of vassal priest, but his role as medium between deities or ancestral spirits and lay supplicants is manifested in various forms in many other societies. Commonly identified as a shaman, the medium may help orchestrate songs, dances and trance sessions to invoke the goodwill of hidden but powerful forces. Such mediums were common in many foraging societies and are undoubtedly of great antiquity in prehistory.

One example of such an exhortative song comes from the Batek Negritos of Malaysia. Like the Ibo of Nigeria, the Batek perform first-fruit ceremonies in thanksgiving for the gift of the fruit. Like the Ibo, they prohibit any cutting of fruit or branches or plucking before ripening.

According to K. Endicott, the Batek visualise their nomadic quest for food as part of a preordained pattern which puts them in touch with the physical and organic world and forest spirits or *hala*, which sometimes take the form of birds or bees. One of these birds, the spider-eater, is credited with having shown humans which fruit could be eaten, raw or cooked, and taught how poisonous types could be made edible. Outside the growing and harvest season fruits are envisaged as heavenly flowers waiting to be rumbled out of the sky and down on to the trees by the thunder god, Gobar.

In addition to harvest ceremonies the Batek have spring songs to entice the *hala* into being generous with their flower showers. Many Batek believe that if they do not thank both the *hala* and Gobar there may be no fruit at all and various punishments may befall them, including being attacked by a *hala* in the shape of a tiger.

Both thanksgiving and enticement ceremonies consist of singing under the fruit tree and burning incense, and among the singers there may be a fruit shaman. The shaman's presence offers the sneak chance of some extra fruit, because at the climax of the repetitive song he may go into trance, sending his shadow-soul to steal some extra flowers. The catch is that if Gobar should notice and kidnap the shadow-soul, the shaman's body back on earth will expire.

This is the praise song for *tawes*, which the Batek believe was the first of all fruits to be created:

> *Tawes, tawes, tawes, tawes*, very red when it is ripe,
> very splayed are its branches, banded is its trunk, it
> sheds its fuzz, red is its waistband, red is its seed,
> exceedingly sweet, I really want to swallow it, I
> really want to eat its seeds, very beautiful is
> its trunk, dense are its branches, very small are
> its leaves, very small are its thorns, very fine are
> its thorns, yellow is its flesh – I love it –

I love it truly because it is exceedingly good –
Exceedingly, exceedingly, want to open, want
to eat my food – Oh-oh-oh-oh
I love *tawes* very much.

No wonder the fruit shaman goes into trance.

5: Tools, Techniques and Time

Prehistory is eerily empty without its tools and the details of technology are a necessary context for understanding the diversity and distribution of people, past and present. Artefacts and techniques took people to places that they would never have entered unaided and as the first Modern humans dispersed over the globe they typically found technical solutions to local problems. Warm clothing, water containers, boats, special foraging techniques and clever uses of fire and smoke permitted the colonisation of otherwise uninhabitable lands, stressful habitats and unreachable continents or islands. In their adaptive responses to those stresses only the fittest survived, but time and a large measure of isolation were needed before changed ways of life found expression in changed sets of genes because technological development is so fast. Many scores of generations had to live and die before artefacts could influence the genetics of their makers, and even then the influence was indirect. Novel inventions still expose humans to challenges but they no longer drag physical adaptation in their wake.

The evolution of our own lineage involved a quickening of the pace of human inventiveness but it was still slow enough to help shape the radiation of Modern populations. Before the emergence of Moderns, their predecessors had already developed various uses for fire, had refined food-getting tools and weapons, had devised structures for comfort or shelter against the elements (such as clothing and huts) and modified the environment with fish and game weirs or annual burn-offs of the vegetation. They may have developed simple marks as abstract communication signals and it is even possible that they anticipated Modern humans in the use of rafts or watercraft.

Some of the harsher risks of life – starvation, predators and extremes of climate – had already been modified, offset or diverted by man-made devices before the arrival of Moderns. Erect and Heidelberg humans had enlarged their toolkits and thereby multiplied their share of resources. It used to be generally assumed that the Acheulean stone tools that preceded those of the Middle Stone Age were the handiwork of Erects but at the present time there is no certainty about *who* was using Acheulean and Middle Stone Age tools, especially during the periods when they overlapped. Erects, Heidelbergs, even wholly Modern people could have been the users of tools that often defy easy categories and exist with little context beyond the places they are found in.

Many archaeologists have devoted much effort to reconstructing how stone tools were used and artefacts have been clustered into kits. One such taxonomy erected five kits, each with between four and twelve tool types (most rather too variable in shape to be identified with certainty). The many different contexts in which tools are found break down into categories such as kill sites or butcheries (which are often near water, ravines, cliffs or sump-holes), temporary camps, base camps, hearths, shelters, middens, pigment mines and (by extrapolation) plant harvesting sites. For the foragers most of these focal points in their daily life would have been landmarks in a well-known territory. Each member would have had a detailed mental map of the area's resources and would have followed traditional routines for its exploitation.

Each kit was thought to have served a different purpose: killing, butchering, processing, shredding and maintenance. Even so, there are difficulties in recognising all the purposes to which tools were put, especially when shapes are rather amorphous. There is a point beyond which attempts at systematic analysis begin to tell us more about our own claims to be obsessive classifiers than they do about prehistoric behaviour. Nonetheless, Middle and Later Stone Age toolkits were real enough and a valuable supplement to a surplus of stone tools is to examine the infinitely richer repertoire of artefacts made by living or recent hunters, foragers and fisherfolk. With the insights they provide it is possible to surmise that Moderns could have been the first to introduce elements of system into the job of making tools.

The term 'Middle Stone Age' indicates a general lightening of tools, with more skilled flaking and careful retouching, especially of cutting edges. These differences imply that such tools were more portable and their finer finish may have made them less readily disposable. There are also early signs of tools being made for specific jobs.

From the time of their emergence onwards it is certain that Modern humans made and used the class of artefacts now labelled as Middle

(left) Agta boys playing in camp.
*(below) Shredding rattan in an Agta camp
(Photo J. Kamminga).*

Stone Age. There is still some uncertainty whether these were invented by Modern humans but it is a very suggestive coincidence that finer and more diverse artefacts appear at about the time and in the same areas as Modern humans. Recent datings tentatively put the Middle Stone Age at between 240,000 and 40,000 BP, which spans all the early development of Modern humans. It is a false ending because there is actually no conclusive break outside Europe, only a continuum that moves by fits and starts but still shows increasing diversity and invention. In some deposits from the upper levels Modern human remains lie among the tools they made, but for most of the very lengthy earlier periods there is little conclusive evidence of who made the artefacts.

Exceptions come from eastern and southern Africa, where not only the identity of the makers but ecological contexts and events can be retrieved. One of the earliest and certainly one of the best preserved skulls of a fully Modern human has been found on the banks of the Omo River, at Kibbish, in a layer that yielded no artefacts but has been dated at 130,000 BP. However, a similar but more fragmentary fossil of similar age in Kanjera, Kenya, was found among typical Middle Stone Age tools.

One of the most consistent and lengthy series of Middle Stone Age tools found so far comes from the Cape, where a coastal cave by the mouth of the Klasies River was occupied from 120,000 BP onwards. Another site in the Cape is situated on an escarpment above the regular route of seasonally migrating game animals. Such movements, mainly by large animals, would have taken place in July–September and in October–November on their return. The ratio of larger migratory prey at this site amounted to 42 per cent of the midden. The rest was made up of smaller animals that could be assumed to be resident around the camp-site. 'Ambush sites' of this sort have been found in Ethiopia and Asia and were used only during the animals' migratory period. Similar exploitation of migrating herds can still be seen in Canada, Tanzania and Sudan, to name but a few instances.

A graphic picture of an African hunt encampment, from about 85,000 years ago comes from the Orange River, where a series of semicircles mark out the blinds and sleeping hollows of a group that probably numbered about twelve people. They built their hides close to a stream and by the entrance to a gorge, probably embedding thorn branches among the rocks. The site would have been strategically placed to ambush animals at the natural mouth of a channel leading into the gorge.

Near Lake Malawi, at about the same period, a single elephant was butchered by a party that clearly divided the animal into three parts. The remains of the elephant formed three clusters, about five metres apart, and each was accompanied by a scatter of forty or more stone flakes. Elephant skin and tissue is very tough and blunts blades very quickly, and one can imagine the chipping and flaking that interrupted the gorging. Since the amount of stone needed to butcher such a large animal

Cleaning a skin.

could not be carried, each hunter would have had to be a very proficient tool-maker and be able to use whatever stone was immediately to hand.

Butchering must be fast because meat soon decays in a hot climate and the build-up of scavengers can be sufficiently great soon to drive a small party off their kill. Prehistoric people must have been less fastidious than we are about rank meat but they would have recognised decay as toxic and known the effect of eating dangerously rotten meat. Speed was of the essence and many large animals must have been cut up into loads and carried back to base-camps, because these sites, unlike the killing ground, have only a few selected bones from large animals and the remains of many smaller, more portable species.

The habitats that Middle Stone Age people were best equipped to handle were those that could be fired and this preference may have been maintained for very long periods of time. Ecosystems where there was no fuel for fire may have been avoided because they required forgoing a prime advantage humans had over other predators. Deserts, arctic tundra and very wet forests were habitable at a later stage and, in every instance, required the invention of specialised subsistence techniques.

Estimates for the earliest Middle Stone Age in India range between 163,000 and 120,000 BP but 144,000 BP may be closer to the mark. The first sign of a break in Acheulean tool-making traditions in the Middle East come from the Jabrud rock shelter in Syria, where small hand axes and numerous angular scrapers suddenly appear dated to about 150,000 BP. Given the close association between Modern man and Middle Stone Age tools it seems likely that the occurrence of this industry marked his first arrival in India. After the initial spread by a small number of founders, there need not have been much significant genetic exchange between Africa and India thereafter. The maintenance of this tradition in both areas over many tens of thousands of years could, therefore, signify that the subsistence techniques of which they were a part were extraordinarily effective. The stability of Middle Stone Age life over such a long period must reflect a very successful formula for living off particularly reliable resources, which are known to have included many large mammal prey. Minor innovations in technology may well have diffused over the western Asiatic land bridge, but the overall picture is one of an equilibrium which did not involve major changes until about 25,000 years ago. Indeed, the Middle Stone Age in India is best represented from 40,000 years onwards.

There are, of course, both temporal and regional variations but typical ranges of tools have been excavated. In a factory site in the Luni Valley, in Rajasthan, Dr Mishra and his colleagues found that 46 per cent of the tools were scrapers, 20 per cent were points, 12 per cent were cleavers or hand axes, 8 per cent flake knives and the remainder were awls or borers and combined scraper/borers. The materials used in India tend to be types of crystalline silica such as agate, jasper or chalcedony. Decorated fragments of ostrich shell (a long-extinct bird in India) are known from Middle and Upper Stone Age deposits, and a small bored shell suggests body ornament. There may have been an aesthetic as well as a functional dimension for prehistoric tools, as has emerged from studies in Australia. The likeness of milky translucent quartzite to animal fat is thought by the Tolngu people of eastern Arnhemland to invest quartzite points with special killing power. This property derives ultimately from the authority of great ancestral spirits.

Most of the Indian artefacts have been found in the drier west and centre. In the moister northeast, Moderns may have made their first accommodation to more heavily forested country. In any event, there would have been a pause in Modern expansion eastwards because their established technology was not adapted for life in moist forests.

Further north there were no savannas in subtropical and temperate China but there were different challenges. As explored in earlier chapters, Modern humans could have encountered more than one precursor in this huge region of high mountains, valleys and plains. The apparent near-absence of Middle Stone Age archaeological finds, in striking contrast to India, implies that it took them a very long time to reach dominance in an already well-settled and highly diversified region.

In the upper cave at Zhoukoutien, near Beijing, the remains of indisputable Moderns are thought to be associated with a mixture of Late and Middle Stone Age-type artefacts. The resemblance of some of the tools with the Upper Stone Age in western Asia implies that these cave deposits were relatively young and, so far, all the Modern remains found in China are thought to be less than 20,000 years old.

To the south, Modern humans also faced well-established but more primitive human societies. The present island of Java had been joined and separated from the mainland at least twice during the million years that it was occupied by Erects. Although artefacts and human remains have yet to be found together on the same living floor, Javan Erects must have made tools but stone was not the central material. Small flakes and longitudinally flaked pebble tools as well as large choppers or 'Oldowan' hand axes have been found but their contexts are virtually unknown.

To some extent the rarity of tools for the period when Moderns first entered the area could reflect a real rarity of people, but it may also imply that preferred littoral sites are all under water. It is out on the islands that most of the evidence for the supposed presence of modern humans has been found but the signs are relatively late and consist of distinctively flaked tools. However, the islands could not have been colonised without boats and their general use is unlikely before about 130,000 years ago. At that time Erect humans were probably still living on Java and possibly scattered elsewhere in the drier parts of Southeast Asia.

Modern humans would have learnt many things for themselves but some of the basic rules for subsistence in this region may well have been learnt from those who preceded them. If Erects had devised efficient ways of using the many resources and materials of the region with little recourse to stone tools, then there would have been less reason for Modern humans to continue using their savanna-designed Middle Stone Age toolkits.

Forest-dwelling subsistence economies survived in several parts of this region until very recently. In the absence of a fossil record they are some guide to how Modern humans first coped when they entered the region. Of course the central ever-wet forests are likely to have repelled all encroachment until much more recent times but their edges and the more seasonal monsoon forests would have been more inviting.

The basic pattern of men hunting while women gathered plants was very widespread. In the Philippines, Nigrito women and children foraged in the forest, using sharp digging sticks to unearth roots as well as grubs, snakes and rodents. They used woven bamboo traps and hand scoops to catch shellfish, prawns and fish in small streambeds. In the Andamans, women wove two sizes of multipurpose baskets, the smaller *toboaga* and the larger *toblaya*; among their skills were pottery and the making of bark cloth, but both these are known to be late introductions. It is the difficulty of separating the more ancient residual elements from such later influences that counsels caution when modern or recent peoples are posed as models for the distant past.

They are, nonetheless, a means of bringing many aspects of prehistory to life. For instance, the Dyirbal forest people of northern Queensland illustrate how marginal or wholly dispensable stone tools can be. Although they used stone axes and stone blades these were not a prominent or essential part of their kit. Spears made of hardened wood were used to hunt for wallabies, possums, echidnas, eels and fish. Throwing sticks were used at dusk in the canopy to knock down fast-flying pigeons, parrots and fruit bats. A highly regarded guild of skilled tree climbers shinned up to the tops of trees to gather canopy foods such as honey, tree kangaroos, opossums, lizards, birds' eggs and fruit. They carried baskets made of bark or creepers on their backs and woven dilly bags hung from the shoulder.

Ingenious traps caught various animals. Small birds were limed with a gluelike sap from fig and Pisonia trees which was spread on prepared branches; woven baskets baited with ant-larvae were lowered into the water to capture freshwater shrimps. Fish were caught with plant fibres, gill nets or on a line baited with worms. The catching device was either the lignified tendril of a local vine, serving as a hook, or naturally barbed shreddings of rattan cane (rattan is commonly called lawyer-vine because of its hooks which are very difficult to disengage).

Animals and plants alike were sliced, split or shredded by ground-edged pieces of snail shell; and long, patient grinding of points and blades in wood, bone or shell was as effective for most purposes as stone equivalents. Judicious scorching with fire can either harden or crumble wood and bone. Thus selective charring allowed throwing sticks to be

Yidinj (neighbours of Dyirbal) stone axe.
Yidinj dilly bags.
Yidinj hand net for fishing and tubular eel
trap of cane.

TOOLS, TECHNIQUES AND TIME

fashioned from flat tree buttresses or spears from surface roots or saplings. Fire was also used to cure, harden and straighten spears. Fire is a primary tool in the forest. Smoke can be used for neutralising bees during honey collection or for flushing game out of various kinds of retreats. Smoking was also the preferred treatment for the very abundant Queensland eels which were caught in rattan eel-pots. Well protected by a slime-producing skin, these eels had to be gutted and lightly scorched on hot coals so that vigorous rubbing on grass would remove the slime coat. Then, placed on a tablelike platform of green sticks and covered with strips of bark, the eels were cooked for two hours over a smoky fire that was contained by a tent of leafy branches.

Other foods such as birds, marsupials and fish were parcelled up in great ginger-leaf wrappings and placed on preheated stones lining a small crater that had been dug (preferably in sand), with large leaves above and below and a sand layer to seal the oven in; the fire was revived and two hours of firing completed the baking. The hot sand was removed with a fan-palm brush and the food was lifted, unwrapped and eaten. This method of baking is very widely practised in many parts of the South Pacific.

Dyirbal dwellings were similar to those of other forest people, including the Mbuti Pygmies of Africa. The building of these huts called *mija* was described by Eric Mjoberg in 1918. The people first selected a suitable site which was cleared and tidied. The women went out in groups and collected rattan cane and masses of palm leaves. The men then assisted with the building of the hut.

First they put up a sturdy framework. A large number of thick saplings taken from smaller forest trees were placed in the earth, in a circle or oval. Sometimes living rooted saplings were simply bent over with their ends together and under each other, so that a sparse frame was formed. They were then bound carefully together with rattan cane split into tough switches. This lattice was filled up with many layers of overlapping rattan fronds, leaves and bunches of rushes or bark when built for the wet season. *Mija* for the windy, cool season were sited in well-sheltered spots. These were smaller, more squat to the ground and thatched in good insulating layers of grass and paperbark, all strapped down with vines and split rattan. The lattice wood inside was hung with baskets, bark blankets and tools, often swinging from wooden hooks that were fashioned from tree crotches or roots. Where people slept a pile of dry grass formed a mattress.

A special rattan basket was used in the processing of one of the Dyirbals' staple foods. A very common riverine tree, the black bean, sheds great numbers of heavy, brown pods containing walnut-sized poisonous seeds. These were first softened by lengthy steaming in an earth oven, then, after thorough dicing, the Dyirbal filled the basket and suspended it in running water for at least a day and a night to leach the bitter toxins out of the seeds. The residue provided a tasteless bulk filler or dough but it had the virtue of being storable. After cooking in ginger leaves it could be wrapped, buried and eaten up to six months later.

Australian tools, techniques and some of the ideas associated with them help illustrate some broader aspects of prehistoric practice and demonstrate many details that are relevant to the overall development of technology. Australia is of special significance because, although both genetic and cultural influences must have filtered in from outside (betrayed by the late arrival of dogs and pottery), a basic matrix of very ancient tools was retained (spears were in general use, bows and arrows absent). This matrix was influenced by local losses, such as boat-building skills in Tasmania, and by gains such as the invention of the boomerang. However, the limited extent of external influences over a period possibly in excess of 60,000 years serves to provide a sort of technological bench mark by which to portray the basic character of early human technology.

One of the most valuable aspects of Australian technology is the insight it provides on how perishable materials may have been used in the past. Australian Aborigines use many natural materials for their unique properties of mechanical strength, durability, flexibility, consistency or shape, most of which are absent from stone. Furthermore, Australian tools have real contexts, not just guessed functions. There are even contexts of individuality and status for their makers and owners. There is a richness of meaning in the barest of Australian subsistence economies that is almost totally absent in archaeological relics.

Because all the equipment that cannot be devised on the spot has to be carried, prehistoric people and modern foragers combine as many functions as possible in one implement or, to invert the priorities, make one tool do a lot of jobs. The premier multipurpose tool is the Australian spear-thrower, or *woomera*. The flat blade of this implement has its spear-

The spear thrower or woomera. *(The force of the throw bends the spear quite sharply as it leaves the woomera.)*

throwing wedge mounted at one end (this is sometimes carved into a formalised animal shape). On the tip there may be a sharp flint embedded in mastic which can cut, bore or chisel as required. At the other end the hardwood blade or its point may help provide the friction to ignite tinder, as well as serve as a club or rhythm stick; the flat can be used as a small palette or miniature mixing platter. Markings on the spear-thrower can be a personal passport, a recording slate, a 'map' or a message board. Similar multipurpose tools would have been widespread throughout prehistory and the spear-thrower itself is likely to be of great antiquity. The *woomera* was effectively the first 'gun'. When it is used properly with a perfectly balanced spear, it can project that missile with great force, steadiness and accuracy over distances of 100 metres.

The principle on which the spear-thrower works is to extend greatly the length and therefore increase the leverage and momentum of the throwing arm. The essence of skill with a thrower is to transfer accuracy of aim and trajectory from the hand out along the lever to the spear's precarious seat in the wedge.

Although boomerangs are known from Africa, they were invented quite independently in Australia by the Aborigines. Boomerangs are an elaboration of throwing sticks. The curve or angle in a boomerang should be 'natural' and cut from a branch or buttress where the grain of the

177

A woomera or spear thrower marked with a
'map' of sixteen waterholes along a 240 km
track of the Totemic rainbow snake (Courtesy
J. Kamminga).

wood and growth direction match the weapon's final shape. Some are much more angular than others but all have one side flat and the other slightly convex. Hunting boomerangs do not return to the thrower and their main requirement is accuracy, velocity and force of impact. They were probably first developed as a means of felling the flocking birds that are such a major feature of Australian ecosystems. Thrown above panicking waterfowl, spinning boomerangs drove the ducks into flying so low that they went into preset nets. This and numerous other descriptions show that boomerangs were used with great dexterity and ingenuity to control behaviour as well as actually hit their prey. The same principle seems to have operated in warfare, where boomerangs might wound, frighten or panic the adversary but spears did the killing. Likewise, a large kangaroo might be turned or maimed by a long-distance throwing stick but it was a spear that finished it off.

A boomerang is made to return by wetting and heating it to help induce a twist so that the ends bend in opposite directions. Aborigines made this type of boomerang for entertainment. Since throwing needs skill it is an ideal vehicle for displays of special dexterity and finesse. The principles of boomerang flight were so well understood that a dropped leaf of the Brigalow acacia, which has the same shape, can be poised by a skilled thrower and then flicked with a finger; in a miniature exposition of aerodynamics, the leaf shoots out only to veer off and return in an arc.

Another artefact that may be significant for the light it throws on prehistoric designs and decoration is the koong–ga*, or message stick. Sometimes described as a 'passport' (because it could serve as a guarantee of good faith in potentially hostile territory), this small decorated shingle was carried by an agent, usually a brother or close friend, on behalf of some well-known man. The latter resorted to a messenger or koong-ga when unable to conduct his own business, such as an exchange of goods or promises at a market or conference. The sender relied wholly on his agent to relay his message verbally, while the stick was his individual signet-mark or brand. The marks cut into the stick varied from rudimentary to elaborate. Any 'meaning' a mark might have had was totally arbitrary in the sense that the maker could simply name identical

---

*Names for artefacts varied from tribe to tribe, even from one dialect to another. Koong-ga was the name for message stick used by the Kalkadoon of Queensland.

Koong'ga or message sticks:
(a) A message from Oorindimindi to 'Billy',
asking for spears.
(b) Summons to a Meeting. Lower section =
the head campsite of the Boinji sender. From
bottom up W = sandhills; N = a sandy creek
and more sandhills. Horizontals = Marion
downs. Verticals = open plains. Cross hatching
= Tediboo or five-mile yard. W = river. X =
meeting place at Warenda.
(c) Two sides of a summons to a dance or
corrobboree. The two lines on the reverse were

translated as 'Quick, hurry up'.
(d) Xs at top and bottom = beard, the diamond
a vulva.
(e) The sender calls for his spears, boomerangs
and shields to be forwarded to the Mitakoodi
river where an ininitiation ceremony is to be
held. The addressee is invited to join in.
(f) A message from 'Sandy' at Carandotta to
'Kangaroo', requesting pituri meal. Spears
and boomerangs available in exchange.
(g) From Mitakoodi, Cloncurry.
(h) From Barclay Downs.

marks as hill, river or camp. This arbitrary naming of simple geometric marks is very widespread in early 'art forms' from all over the world and has nothing to do with representation. The marks were described in Australian pidgin as being 'flash' to make the stick 'pretty fellow'. It could be cut from any woody material from a sliver of stink wattle to a whittled portion of rattan the name *koong-ga* also applied to any type of wood.

Rattan is an extremely common plant throughout the forests and swamps of Southeast Asia and tropical Australasia. In addition to its edible shoots and berries, its leaves, ribs and canes provide a flexible, very tough material which can be slivered, bent, split, shaped, bound and

Noosing the head of a shot monkey through a slit in its tail converts the prey into a ready made carrier bag, Babu a Baka from Cameroon (Photo Lisa Silcock).

A cane basket (for storing grain) in course of construction. This mode of basket weaving is very widespread (Kigezi, Uganda).

then fixed by being dried out, either naturally or over a fire. Rattan makes all manner of basketry and was widely used to provide handles, because a split green cane could be looped round a block of wood, stone or bone (ideally resting in a rubbed, scorched or chiselled groove or waist), bound all over and dried to make a club, axe or pick.

There are ample precedents to be seen in nature where tropical forest plants wrap around, strangle, topple or split other plants, lumps of earth or stones. It is not uncommon to encounter stones firmly encased in roots or stems of vines and figs. One ingenious Aboriginal technique is to train a living liana around a prepared axe or adze head and leave it to become embedded before 'harvesting' the naturally hafted tool.

As one of the commonest and most accessible materials with many different properties, from stringiness to flakiness or flexibility to rigidity, bark could serve as a hut door, a temporary cradle or bowl, or provide the material for a cloth or blanket. Some barks also provided that most fundamental of materials, string or cord.

There are many creepers and lianas that are so tough and flexible that they are in a real sense ready-made string. Other materials can be used for this purpose. Even today hunters empty a length of gut to tie their quarry up for transport back to camp. A split tail or loop of skin can make good shoulder straps and strips of raw hide or sinews possibly served as tool bindings for Erect as well as Modern humans.

Like many other elements of technology, the use of strands of fibre has ample demonstrations in the natural world. The fact that foragers make direct connections between natural structures and man-made equivalents is shown in the language of many isolated tribes and peoples. Birds' nests, for example, have the same name as woven dilly bags in Papua. Weaver birds, tailor birds and bower birds are only the more sophisticated of innumerable birds that manipulate strands of natural material to make nests. Many rodents cut and carry grass or twigs which are webbed into nests. Spiders, especially the large tropical species, make conspicuous and extensive trap systems that involve various principles of angling, netting and trapping. Of plants, rattans are the most entangling and several fishing societies, like the Dyirbal, used their barbules to snag fish. Rattans are effectively palms that have become lianas.

Palms are one of the most obviously useful plant groups in tropical forests. Palm fronds would have provided natural shelter even as their sometimes edible dates were another incentive to seek them out. They are also typically riverine and coastal plants. Palm fibres are ostentatiously shredded by nesting birds and loose fronds are an easily available material for shelter from rain, wind, sun or to lay on the ground. In many species the long, double row of leaflets invites interleaving to make the leaf surface into a single plane rather than a ragged frond. This simple intervention, widely known in eastern Africa as *mkuti*, is likely to represent the 'beginning' of matting and perhaps of weaving. Many solid leaves have temporary uses, but the durability and toughness of palm leaves would have commended them for the manufacture of all manner of mats, baskets, roofs and screens.

Huts, shelters, fences and fence traps require abundant lashings. Snares, traps and nets are major uses of high-quality cordage. Bows, arrows and harpoons need very durable bindings, bowstrings and tail-strings. Weapons or other valuable objects frequently need to be suspended from a shoulder or hung from a beam or branch in a string bag or from a leash. Likewise calabashes or other containers for water, honey or blood need string cradles or yoke bindings. Climbing after beehives often requires the improvisation of ropes. Captured or killed animals need to be tied or bundled up or their hides stretched by thongs or cords.

In east African savannas the fibres preferred by contemporary foragers are likely to include those chosen by their prehistoric forebears. The premier choices are bark fibres from the baobab *Adansonia*, and wild fig,

181

*Palm leaves are readily woven into panniers
and rucksacks.*

Ficus petersee. The bark of several species of Acacias and Miombo, *Brachystegia* also provide good cordage.

Among the many fibres available for the making of ropes in rain forests are wild banana (*Musa textilis*), which is easily shredded and, when dry, has very fine, long strands. Various types of hibiscus, hemp and a multitude of forest vines and tree barks widen the choice. Short, small fibres of animal or plant origin can be easily plaited into a thread by rolling them on the thigh with one hand, while feeding fresh fibres, a few at a time, into the twist. Rolling two threads into a two-ply string can serve some purposes, while further plaiting can generate any thickness up to strong rope.

The webs of many tropical spiders can be rolled into very strong twine or used as a sort of self-adhesive binding material (a sort of prehistoric Scotch tape). Mallanpara Australian Aboriginal children used to fish for tiddlers with a very short line-cum-bait made from spider webs. Their method showed that the principle used by one species to catch its own prey could be taken over and restructured to catch a completely different prey.

Many spiders first peg out a structure of smooth elastic lines. On this radial plane they lay down a single continuous spiral of sticky web on

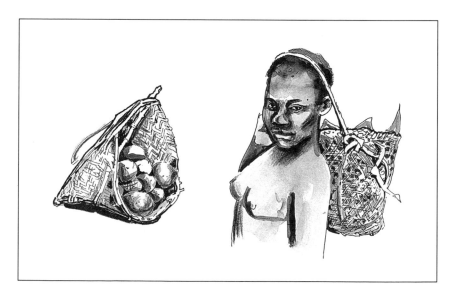

(left) *Plaiting string by rolling small fibres.*
(right) Weaving palms and other fibres to
make rope.

TOOLS, TECHNIQUES AND TIME

which their insect prey are caught. The Mallanpara first caught and killed the spider; then, having preserved the web sac in its abdomen, they proceeded patiently to unravel the web. With the tip of their rod they lifted the sticky web off its non-sticky guy ropes, moving the twirling rod tip, spoke by spoke, around the ever-decreasing spiral. In this way dozens of little loops accumulated into a single dangling mass at the end of the rod. While each loop added its strength to the free line, a twirl of the rod bound an ever tighter attachment on to the tip. By the time a web had been unwound there was a 'line' on the end of the 'rod' that was composed of innumerable sticky loops, and this was danced, or bobbed on the water surface to attract small fish. Every bite caused the fish's jaws to get both entangled and glued to the line and thus caught by the fisher. The line was re-baited and re-glued by dipping its tip in the spider's web sac and the method was said to be a very rapid (if elaborate) way of catching very small fish.

Its significance lies not only in the ingenious diversion and adaptation of a naturally observed principle but in the near certainty that it was invented by children (although the surgeon and anthropologist Walter Roth, who first described it, said that adult men also went tiddler fishing). If the invention of such technology was 'child's play', playfulness,

183

juvenile intelligence and manual dexterity may all be central to the development of Modern human technology.

A second example of fishing with spider webs raises still further questions of how prehistoric people might have observed phenomena and then refined and systematised them to their own ends.

Forested foreshores in the tropics are often literally webbed from end to end every night by tens of thousands of large spiders. At dawn, when the birds begin to fly, the air is thick with gossamer as broken threads and tangles waft out to sea. Garfish, among other surface feeders, are quick to snatch at the debris, and fisherfolk from the tiny island of Iwa, east of New Guinea, must have noticed this phenomenon because they long ago devised a unique form of fishing with kites on which spider-web fishing lines dangled from the tail. The kites were made from large but very light leaf skeletons reinforced with fine twigs. They were controlled by a fisherman in a canoe with strings about two metres long. The freshly gathered spider webs were hung in skeins from a twine tail and, as they skipped over the surface of the sea, attracted garfish which got their sharp teeth snagged when they bit. The fisherman then drew in his kite, disentangled the fish and continued to fly his kite and catch further fish.

The structure and controlling mechanism of the kite were almost certainly devised in complete isolation on this remote island. The perfect grasp of aerodynamic principles and control via fine strings implies that the observant eyes, ingenious minds and dextrous fingers of prehistoric people must have devised many interesting but now lost ways of catching prey. At the root of this ingenuity is a real feeling for string and its properties.

String is an essential precursor to a vast number of traps and snares. Fish traps in particular tend to range from rigid woven baskets and webbed barriers to various types of more flexible netting (in which rattan may serve as a semi-rigid reinforcement or boundary). String, fish-traps and hunting for food in the water might also have been essential precursors of boats and rafts – or were they?

The most rudimentary form of raft or canoe must have been invented or stumbled upon many times, taken many forms and served quite different purposes. It is useful to consider the dilemma faced by wholly terrestrial people when they had to ford rivers or flood-waters. Predatory crocodiles would have been a nearly universal hazard and a serious disincentive to systematic foraging in or near tropical rivers and lakes. There would have been 'squid techniques' of diving to the river bottom and stirring up a screen of mud – but that required an ability to swim that might have been rare. Alternatively, people could have gathered in a tight gang and charged across with the maximum of splashing and shouting. Again, the ability to swim and the presence of companions would have been necessary. A safer and subtler strategy may have been practised very widely in prehistoric times and could have been one precursor of boats. It survived up to the turn of the century in Queensland where a scatter of cut mangrove logs lying about the banks of the Mitchell River caught the observant eye of Walter Roth.

These two-metre lengths of very light, buoyant wood were 'ridden' across the river, paddled by the Aborigines, who would lie stretched out on the log. Roth remarked that these ferry logs, so handily left on the bank,

> might serve the purpose of protective mimicry from the attacks of crocodiles, which literally swarm in these waters. The thinner end of the float, which projects behind after the nature of a tail, giving the swimmer all the appearance, at no considerable distance, of one of these saurians; that the natives here have but little dread of these creatures may be gauged from the fact that on the occasion of a visit of the government ketch 'Melbider' to the Mitchell River, eleven crocodiles were to be seen at one and the same time from the vessel's deck.

Roth also recorded larger floating logs being used by two or more travellers to cross stretches of sea as much as ten kilometres wide that lay between the Keppel Islands and the mainland. Here the Aborigines swam hanging on to each other in a zigzag formation while the leader guided the log float with one hand. Apart from their role as safety floats, such logs allowed long-distance swimmers to rest and alternate in taking the lead; but there were other roles for floats. The well-known Aboriginal trough or *koolamon* was a hollowed-out longitudinal food bowl that could be floated across rivers while its owner swam or waded behind. Yet another device for crossing Queensland rivers was found on the Tully, where four or five logs were tied together in a crude raft that was in temporary use during floods.

The materials most suitable for devising a float differ greatly from place

Log foat or 'Keppel island water taxi'.

to place. For example, boys fishing on coastal rivers in northern Papua New Guinea make rafts by aligning a dozen or more lengths of very large bamboo stems and lashing them down above and below on crossbars. This 'floating door' type of construction is unable to resist much flexion or diagonal stress and so is used only on open rivers on calm days. Here there is effectively no boundary between a toy and a technique of subsistence, specially geared to juvenile energies and education. Much stronger and better-made bamboo rafts are widely used on the coasts and rivers of China and Indonesia.

On Lake Turkana in Kenya, a smaller number of long straight trunks of lightweight ambatch trees or *Phoenix* palms are lashed together with short crosspieces to make a sturdy raft. Nets are set and lifted and fish speared from this platform which is quite resistant to being buffeted by waves.

On the lower stretches of the Tigris and Euphrates rivers in Mesopotamia, where reeds were virtually the only local material that would float, reeds were tied and re-tied in long fasces to make what was in effect a boat-shaped raft. The dependence of people on local material is perhaps best shown by the *shashas* of the Batinah coast of Oman. Here fishermen make this one-man boat-raft from date-palm fronds which have been dried and stripped of their leaflets. Tied together at both ends with palm-leaf rope, the *shasha* floats for an hour or two before becoming water-logged; long enough for the fisherman to visit his nets and fish traps. Hauled out on to the beach it will be dry enough for another trip on the following day. Groves of date palms line the Batinah foreshore and are the only large trees to grow there.

Fishing raft, Coromandel coast, India (Photo Belinda Breeden).

In all these instances the rafts are used for short periods for specific purposes, in the hands of a limited class of persons. In every case they use a dominant plant or the only plant material to hand.

The dangers of rafts are highlighted by the Aborigines in the Gulf of Carpentaria, who used crude mangrove-pole rafts until this century. The fishermen of Bentinck Island would regularly visit reefs and offshore islands for the special resources they offered, 'riding tides' to do so. The risks were recognised as very high and more than half the men attempting tide-ridings of more than 14 km were likely to be drowned. Even more dangerous tide-riding on rafts used to be practised by Aborigines of the Buccaneer archipelago to cross the Sunday Straits in northwestern Australia. Here high waters, bottled up in the 3,800 sq km of King Sound, rush out through a narrow neck and back again with every tide. Skill in these circumstances is subsidiary to luck and drownings were frequent for the fishermen of Cockatoo and other islands.

The greatest diversity of boat types is found in southeast Asia which at least one authority on boats has called the 'Indonesian centre of boat diversity'. This is where the invention of various types of watercraft first shaped human destiny. Whether they developed boats from scratch or

picked up some rudiments from Javan Erects is immaterial to the fact that it was Moderns that first made water vessels the 'open sesame' of vast new resources. Such a revolution could not happen all at once. To have a boat was one thing, but what could be harvested from it and how far it could go depended upon a series of subsidiary and later inventions, such as fishing devices and sails.

At the most immediate level, the possession of boats undoubtedly changed the definition of seashores from being the outer edge of habitable territory to being the central core of existence. Thanks to their boats a marginal population expanded very greatly, both in numbers and territory, to become a highly distinctive and specialised branch of the Modern human family. Developed in the relatively limited context of large rivers, boats took the waterside economy out of the estuaries, along the seashore and then out to the islands. The first incentive was an abundant and diverse source of food, previously unexploited. Combined with the monopoly of potentially vast new island territories, this gave huge demographic advantages to the boat people and there would have been a rapid expansion of numbers throughout island Southeast Asia.

The 'Indonesian centre of boat diversity' may have been diverse from the start. If the people were as adaptable as I contend they were, the materials that were locally available would soon have led to local differences in boat construction. As the *shashas* and the Mesopotamian reed boats demonstrate, the boundary between raft and boat can be blurred when the craft is of crude construction. Nonetheless, several essential differences would have emerged very early on. Canoes, whether of bark or wood, *contain* the boatman, his equipment and catch. Unless surrounded by a border, rafts *support* him; the rougher the water, the more important this difference could have become, because of the need to tie things down to prevent them from slipping off. A more significant difference was the greater ease with which a canoe could be manoeuvred and a considerable difference in the speed of the two types of craft

The estuaries and rivers where watercraft were first developed are frequently clogged up with weeds, reeds, tangles of flood debris and stranded tree trunks. Since a clean fallen trunk is the easiest thing to manoeuvre and push through such an obstacle course, I think it more probable that canoes were the main vehicle for human expansion in Southeast Asia.

Local availability limits the choice of canoe trees, thus Roth noted the tall mangrove (*Sonneratia acida*) or cheesewood (*Alstonia*) being used on the Endeavour River, grey teak (*Gmelina*) and white mahogany (*Canarium*) at Cape Bedford and silk cotton trees (*Bombax*) on the Batavia River in Queensland. Very widespread in the Old World tropics is the Karaya (*Sterculia*), which was particularly suited to the early canoemaker. Many *Sterculia* species are fast-growing, riverine, forest-edge and glade colonists; they send up a long, clear and symmetrical bole up to 30 m tall and one or two metres wide; their bark is often smooth and pale. Since colonising trees tend to be short-lived and because some *Sterculia* species prefer to grow on riverbanks, they become casualties of floods, so that river-borne *Sterculia* trunks would have been familiar to prehistoric hunters. Their heartwood is soft and rots easily, whereas the cylinder around this central pithy core is harder and more durable. Fungus and insects often attack hollow logs on their damp side and it is not uncommon to find natural 'dugouts' in the forest, wholly made by natural forces. During floods, such 'opened' hollow logs are among the debris that floats downstream.

Quite apart from the shelter that a large floating log can afford against crocodiles or fatigue, there is its role as a pontoon for food and implements. Foragers in shallow water have to carry what they find or throw it on to a bank. When several people are collecting food in any quantity, such as molluscs, fish, turtles, flood victims, the squabs of water birds or lily roots, floating containers for their booty become real assets. This is among several opportunistic uses for naturally occurring 'dugout' logs. On a more sustained basis both dugouts and rafts would have been used more frequently as platforms for finding, trapping or spearing prey. The dugout would have had greater manoeuvrability in this respect, and once its benefits had become appreciated, living trees rather than already dead ones might have been sought out for their trunks.

Sterculia trees and palm trunks are still used for dugouts in widely separate parts of the world. In the Andaman Islands, *Sterculias* were recently among half a dozen tree species chosen for making dugout canoes. There are, of course, many other trees with similar properties, notably species that have pithy cores and very hard outer cylinders. The more rudimentary the tools the more the craftsman is dependent on being helped along by the natural properties of his material. Instead of the carver having a preconceived notion or model in his mind, the early canoe-digger may very well have had differences in wood texture as his primary guide in shaping the canoe. Instead of the model of a canoe existing in the mind of the tree-hunter, the canoe existed within the tree much as a ration of meat existed within the skin and frame of a prey animal.

After felling the tree and removing the bark the stem was hollowed out. At prow and stern the pith would have had to be left intact and any cracks or leaks caulked with clay, wax or resin. The broader lower part of the tree trunk provided the basis for a wide plate on the prow of the canoe. This was the standing platform for the harpoonist, while a smaller flat seat on the stern seated the helmsman. Once roughly shaped, the walls of the canoe are thinned from the inside to make the canoe manageable for one or two people. In the Andamans a very simple outrigger was then added to give the boat greater stability in the sea. Outriggers are essentially a sea, lake and open-estuary refinement.

Cylindrical dugouts spin upside down at the least rotary movement so fishers paddling or poling them are very proud of their ability to keep the canoe moving fast and steadily. They achieve this through a perfect extension of body balance down to the canoe's imaginary centre line, so that any unbalancing movement is offset by a small muscular counter-action. Such skills would have been mastered by the first canoeists but capsizing would have been normal and frequent.

A special virtue of dugouts would have been their buoyancy, which-ever way up they were. Bark canoes are often thought to be a separate

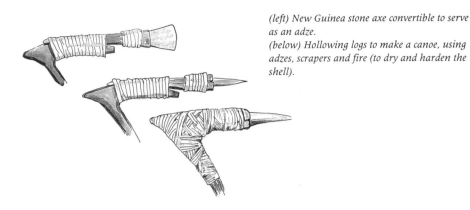

(left) New Guinea stone axe convertible to serve as an adze.
(below) Hollowing logs to make a canoe, using adzes, scrapers and fire (to dry and harden the shell).

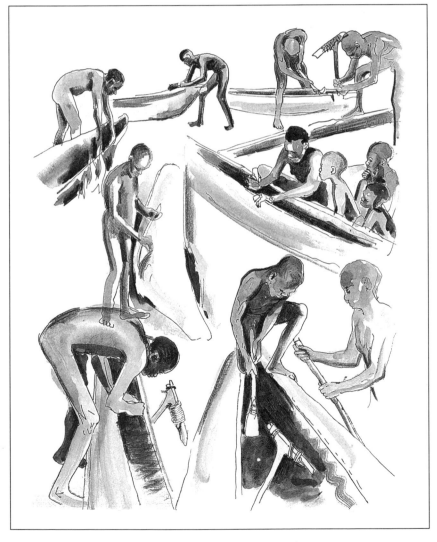

category of boat, yet they are, in effect, another variant of the tree trunk's outer core. Bark canoes have also been thought to be more primitive than dugouts, because they are apparently much less trouble to make. This is to ignore the ecological context and technical traditions in which a local artefact develops.

An awareness of the potential in materials is often developed by previous use for other purposes. That and local availability would have helped decide what materials and techniques were used; also where, how and for what purpose artefacts were made. If bark shelters, bark eating vessels and bark string were the norm and most of the common trees were solid-centred, the labour of excavating hard trunks with quickly blunted axes and adzes would have been inconceivable if a long tube of bark could be modified to serve as a canoe with much less time, trouble and labour.

There were three types of bark canoe in Australia. The simplest Murray canoes were very precarious scoops made from a single arched piece of eucalyptus bark, partially held open by lumps of clay fore and aft. In the second type three or four spacers, short stakes of wood, held the canoe open. Once dry, the canoe would carry one or two occupants. More complex were the sewn canoes. Here the entire bark from a regular

straight bole of acacia was taken, measuring four or five metres. The ends were sewn together to form prow and stern. Two long thin withies were bound to the margins of the bark to form the gunwales and a series of short spacers held the sides open. This type of canoe could hold three or four people.

The fact that bark canoes were the most widespread type of boat in Australia led to the proposition that the first colonists must have come in them. That idea was linked with the concept of a chance landfall by very primitive vagrants. It is more likely that seagoing canoes were in the process of being developed or were already in use at the time of the Sunda-Sahul crossing and that event was the inevitable product of expansion by a fast-growing population. In this conception the Australian coasts were merely a southeastern landfall close to a huge, newly colonised area with a fast-increasing population. The economic base for that population had first been provided by a dense network of short tropical rivers. Then, with the modification of river boats, a marine ecology and technology could begin.

If it is assumed that a large population of boat-and-raft-using strandlopers had built up in island Asia well before Australia was colonised, it is also likely that boat-making techniques had begun to diversify in response to local materials and traditions. Rafts, bark and dugout canoes occurred in various parts of Indonesia and Australia up to the twentieth century. The types most likely to have made the first crossings would have been of the materials, and made according to the techniques, then current in Timor or its neighbouring islands. This southeastern end of the island archipelago is appreciably drier than most of Indonesia, so the choice of trees may have been somewhat restricted. Nonetheless, there would have been ample choice of both bark-bearing and hollow-making trees to choose from.

Although beachcombing is known to have supplemented the diets of many prehistoric peoples, and molluscs in particular may have been essential to survival in many localities, it was boats that changed the way the sea was exploited. What would have confined canoes to rivers was their instability, and the solution to this shortcoming was the outrigger.

Outriggers can be looked at in a variety of ways. The simplest is to imagine how the punter might have stabilised his canoe while engaged in some other activity – he would have laid his pole across the gunwales.

That was one potential beginning. Another is to see the outrigger as a sort of degenerate second canoe. When the pirates of northern Madagascar raided the Comores and eastern Africa (between 1770 and 1820), they used to lash their dugouts together to make them more seaworthy. The great 'double canoes' of the South Pacific deliberately combined two (and sometimes more) canoes in order to carry more cargo, booty or 'marines' on the broad platform that held them together. They were, in effect, the first catamarans.

A third way of viewing outriggers is as a raft that has been opened up. In the Solomon Islands, bamboos used to be lashed together to form two bundles and these served as outriggers for a central dugout. They were held in place by three cross-booms and the resulting craft represented a sort of hybrid between raft and canoe. However, typing and categorising primitive boats in this way is less productive than seeking out the technical problems and ecological constraints that are likely to have operated on boat-builders and influenced their solutions.

The building of a single outrigger on an Andaman canoe does not involve anything that smacks of a raft or of another canoe, it is a simple but clever device for greater stability. The walls of the canoe are perforated on both sides close to the gunwales and the booms – fine, even poles of a very tough, flexible wood such as mangrove – are inserted at right angles to the boat. The pointed ends of these stakes project out a few centimetres on the port (left) and about a metre to starboard. (The port stubs could be conceived as vestiges of the less robust double outrigger but abandoning the second rig would have been a very early lesson in economy and mechanics.) The larger the canoe, the more booms, from a minimum of three up to nine in the longest vessels. Where the booms pass through the hull they are tightly bound on both sides with slivers of rattan, which wrap all the section amidships.

The outrigger float is a very roughly shaped but straight, lightweight plank, broader at the front than behind. It is pierced when absolutely dry by a series of short, pointed sticks made of a similar wood to the booms, hard and resilient. Designed to lie flat on the surface of the water, the float is attached indirectly to the ends of the booms via the sticks, three to each boom. Each set of three sticks pierces the float across its blade; the central one rises straight to near the end of the boom and at right angles to float and boom. The other two sticks pierce the float close to its

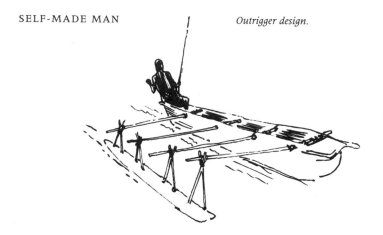

Outrigger design.

margins, and in the same plane as the booms, and rise to converge with the third stick. All three are closely bound together near the end of the boom with split rattan. The rattan also loops down round the float, tying float to boom with the sticks as spacers. (This space approximates to the distance between water level and gunwale hole on the side of the canoe.) In this way the outrigger floats on the surface while the hull rides upright and balanced. The short struts become tightly embedded in the float, because a soft light wood is chosen that swells when immersed in water. Canoes of about five metres can carry three persons (three booms), whereas a nine-metre canoe with up to nine booms can carry heavy loads and about six people.

These lightweight canoes can overturn in rough seas but are easily righted and quickly bailed out with a nautilus or coconut shell. Inevitably canoes ship a lot of water and one person is often bailing full time. The susceptibility of canoes to wash coming in at the bows has been solved in many areas by attaching tapered planks on the bows to increase the freeboard; these are called the wash strake. There have been tentative suggestions by specialists in the history of boat technology that single outriggers might have been one of the more primitive types of seagoing vessels. On the basis of distribution and design details, Sulawesi has been suggested as their centre of diffusion.

If an unadorned dugout log is the primary form from which all subsequent elaborations derive, the single outrigger offers the easiest and most economic modification it is possible to devise. The addition of a float attached by a few booms can be made with virtually no specialised tools and the design has several special advantages for a one- or two-person fishing team. Righting a capsize is much easier, quicker and safer for a canoe with one outrigger than with two. Fishing, spearing, boarding and loading are all easier and less encumbered with one side open. The making and trimming of a single outrigger is easier and repairs are less complicated and less frequent than in a craft with more components to absorb stresses. All in all, the single outrigger canoe remains the main candidate for mankind's earliest ventures in seafaring and possibly provided the means of colonising Australia.

The coasts and larger river systems would have been primary routes for the colonisation of Australia. This was first stressed by Sandra Bowdler, who also pointed out that large-game hunting would not have been important for these wide-spectrum gatherers. Those in the best areas could have been sedentary and achieved relatively large population densities. In support of this conservatism she points to the surprising uniformity and little-changing traditions of Australian stone industries.

The early colonists of Australia probably retained a coastal economy for many generations, yet their descendants eventually became a predominantly continental population. Technologies are usually affected by prolonged isolation and commonly lose many refinements, especially those demanding 'unnecessary labour'. Climatic changes or shifts into different ecosystems also alter the choice of materials and the contexts in which they are used.

When shore dwellers are peripheral, a much larger continental population in the hinterland or interior will eventually be the main source of cultural traits, knowledge and skills. Only on the northernmost edges of Australia were fishing communities in touch (albeit sporadically) with large and populous seagoing cultures. Down in Tasmania, rising sea levels stranded a population of Aborigines for over 8000 years, and it is significant that their watercraft became little more than soggy bark rafts and that various fishing skills, hafting of tools and the use of boomerangs disappeared in the course of their long isolation. The practices and artefacts of more recent Australian Aborigines can therefore be an unreliable guide to those of their pioneering ancestors. In particular, an early shift from a strictly coastal to an inland or continental existence affords no help in determining what sort of craft brought them in the first place.

To seek a hint of what level of technology was operating at the time of that event one should turn to the middens that mark the first human arrivals in New Ireland. The earliest of these have been dated to 32,000

BP and they suggest that most of the food was collected within the intertidal zone. Boats must have allowed people to arrive in the first place, but it would appear that they had not yet become part of a truly pelagic economy, most notably because no fish bones are found. A strong inference is that the normal use of boats and their range was still limited to the human muscle power needed to paddle them and that oceanic fishing techniques were still poorly developed or even absent. This changes at about 20,000 BP when fish bones in a midden at Matemkupkum provide the earliest evidence yet found for marine fish as a major human prey. Bone or shellfish hooks (items that should have survived in the middens) were absent and the excavators concluded that spearing, poisoning, trapping and netting were the most likely methods of capture.

Fishing nets are totally absent from the archaeological record. Early fisherfolk are unlikely to have left large nets out to catch fish passively, because nets cost great effort to make and would often get damaged by other large fish-eaters, so recent fishing practices become an essential guide.

In 1897, Walter Roth published detailed descriptions of fishing by Aborigines in Queensland. The smallest type of net measured about a metre high and about half as wide and was bounded by two curved sticks that could be swept together like the margins of a purse. The single fisherman waded into a likely stretch of river, attempting to clap his purse sticks around any fish that might swim towards him. Another type of net was operated by two or three boys who waded out to form a short line with their nets. These comprised two rectangular frames, each measuring about 3 × 1 m and containing a stretched net with a 6 cm mesh. Once the boys had positioned themselves, with a net held down in the water with each hand, their companions converged on them with much splashing to drive fish into the nets. A third method involved two men in deep water using a long fence net of 15 to 30 m which was woven in a local flax. Each end was attached to a pole and the two men slipped into the water and, keeping close to the bank, stretched the net taut between them. Then one swam out treading water until he was opposite his companion who then set off in the same direction. After 50 m the two swimmers met and drew the net, having swept the upper layer of about 1,000 sq m of river. Often the lower net margin was held down by the swimmers' big toes. In the same region, the fish in shallow water holes

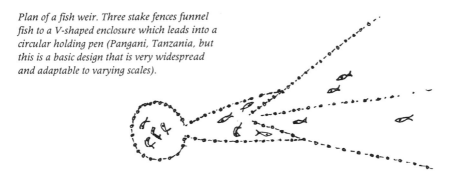

Plan of a fish weir. Three stake fences funnel fish to a V-shaped enclosure which leads into a circular holding pen (Pangani, Tanzania, but this is a basic design that is very widespread and adaptable to varying scales).

would be concentrated by the women forming up behind a wall of branches and grass tussocks which were pushed forward until the fish were confined to a corner where they could be speared, a technique still used in several parts of the world today.

In the Bulia district, the flood waters of small rivers were exploited by building fairly permanent (and regularly maintained) rock weirs. Breaks in the weir went over rock platforms, which had side walls and were covered with a dense mat of leaves or grass, or webbed with a small net after a flood. Fish going over the platform became entangled in either the grass or the net, where attendants could seize them and dispatch them with a bite in the head.

The general principle of weirs is to funnel fish coming downstream into some kind of an open corral that retains the fish but allows the water through. The main risk here is clogging with debris and such traps cannot be left unattended for very long.

For fish working upstream or coming in on a rising tide the weirs and funnels to catch them must face the opposite direction and lead into further funnels, mazes or enclosures that will concentrate the fish. In some of the more extensive estuarine fence traps and weirs, virtually all the fish that fail to swim over the top of the weir while water is high will find themselves trapped by falling water. Here two fishermen must be in attendance as the tide falls to avoid losing their catch to piscivorous birds. Local populations of bottom-dwelling fish could have been greatly reduced by such methods but it is conceivable that this may have increased the invertebrate fauna and thus continued to favour low-tide beachcombing by the strandloper community.

An interesting example of coastal people exploiting and manipulating the behaviour of marine predators in order to share their prey was recorded more than a hundred years ago in the bay where Brisbane now stands. The event was presumably seasonal, when scouts would mount headlands and watch for signs of fish schooling close inshore. The presence of the fish was attended by pods of dolphins, which were

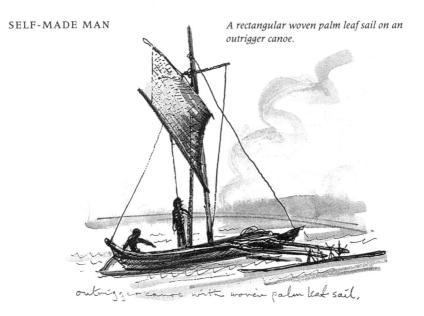

A rectangular woven palm leaf sail on an outrigger canoe.

outrigger canoe with woven palm leaf sail,

sufficiently intimately known to the Aborigines for individuals to have been given names. The dolphins in turn had learnt to respond to human signals. When the water immediately inshore of the fish schools was slapped with wooden batons the dolphins quickly converged on the spot, driving the fish into the shallows where the men speared them and flung their catches ashore. Once the frenzy was over and the fish dispersed the dolphins were hand-fed some of the catch. There are other tales of fisherfolk and dolphins cooperating from Proper Bay in South Australia and in southern Mauretania. A rather larger alliance formed near Green Cape off the southeastern coast of Australia. Here a pod of killer whales learnt to benefit from the whalers harpooning humpbacked whales out of the station in Twofold Bay. Every member of this pod was known individually to the whalers who gave them pet names – Old Tom, Cooper, Hookey, Humpy, Jackson and Typee. Between 1842 and about 1920 the killer whales took to patrolling offshore and would close in on passing humpbacks, then harass and drive them right into Twofold and Leatherjacket bays. Here the whalers, saved all the trouble of scouting, would scramble into their boats, quickly harpoon one or two corralled humpbacks and reward the killer whales with the tongues and lips of their joint quarry.

Such associations have picturesque appeal but they also illustrate an enduring trait in all broad-spectrum carnivores and scavengers and that is the ability to notice and take advantage of any temporary event that offers a free hand-out. Since several different species of meat-eaters are

A woven fibre boom lug sail as used in Papua New Guinea.

nearly always drawn to the same place when these events take place, the faster, more observant species quickly learn to use one another for cues or to divert the fruits of another species' special skills to themselves. I have seen a fish extracted by the probing bill of an ibis seized by a stork which, in turn, lost it to a fish eagle; kites and herons tried to get in on the act too.

The expansion of shoreline foraging to open-water fishing has many implications, among them, sailing. The use of sails is not a necessary adjunct to the catching of sea fish, but sailing would have revolutionised the ease with which prime fishing grounds could be reached and would have substantially enlarged the diet and catchment zones of fishing communities.

In any event, the invention of sailing is unlikely to have been long delayed once people had begun to rely upon harvesting sea fish. It can be supposed that fisherfolk, continually exposed to the vagaries of currents and ocean weather, would eventually turn the flow of wind and water to their advantage. They would have known that it takes no more than a standing figure to drive a boat along at quite a clip when the wind happens to be right.

The most likely sail structure would have been a rectangular mat of woven palm leaf incorporating one or more fronds of *mkuti* and possibly using the rib itself as a boom that was attached high on a pole or foremast. Such a primitive structure could still be called a 'boom lugsail' in modern sailing parlance.

Types of sail.
From left: Boom lug; Trapeze boom lug;
Mediterranean lug; Crane sprit; Double sprit;
Common sprit; Oceanic sprit + frontal view.

Again, Southeast Asia is obviously where all sailing techniques were explored more intensively. Once started, various experiments in rigging structures and techniques would have followed (some of these are illustrated). The likelihood that open-sea sailing and sea fishing were not yet general in the South Pacific by 20,000 BP suggests that estuarine and intertidal ecosystems had gone on being the mainstay of island life for many tens of thousands of years. The absence of sea fish is one of the few respects in which the New Ireland coastal diet of 32,000 years ago differed from that of a century ago. Shellfish, crustaceans, turtles, crocodiles, frogs, birds, fruit bats and other mammals and, of course, plants have remained perennial staples throughout the inhabited South Pacific. Supplements from land-based hunting and gathering diminished only at a very late stage, but the special skills and adaptability required to harvest a large proportion of the diet from the sea served to make such societies very distinctive from the beginning.

On the Southeast Asian mainland, a tropical culture that remained based on streams and rivers might have steadily diverged from its island and coastal descendants but so far it has failed to reveal itself in the archaeological record. A very late excursion into the rainforest's northern margins about 10,000 BP is known as Hoa Binhian (named after a Vietnamese village). The archaeologist Peter Bellwood considers that the sustained absence of sites within the wet rainforest belt is a true reflection of their avoidance by early Moderns. He thinks the Hoa Binhian culture could represent the arrival (possibly from a continental source) of some technical innovation such as bows and arrows or blowpipes.

The flakes, pebble tools and scrapers that have been found in Southeast Asian sites were probably used mainly on wood and perhaps on

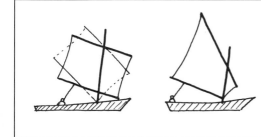

roots, tubers and bamboo. In particular, some flaked calcite blades found in Spirit Cave in northwest Thailand have curved wear on them that suggests that they had been used on hard circular shafts such as bamboo. Bamboo charcoal was also found at these levels of Spirit Cave.

The importance of bamboo in present-day south Asian cultures has encouraged the archaeologist Geoffrey Pope and others to visualise a Southeast Asian culture that relied on bamboo and wood rather than stone, and they project this back to the earliest periods of human occupation. While stone tools may indeed have been dispensable, Bellwood's contention that the ever wet equatorial rainforests were only penetrated along their fringes and coastal margins (and then mostly during drier periods) is probably correct.

In this respect the Southeast Asian archipelago makes a striking contrast with the Indian subcontinent, much of which remained prime human habitat throughout the last 150,000 years.

In Saurashtra, at the northeastern corner of the Arabian Sea, a large number of tools, mainly scrapers and combined scraper-borers, have been retrieved from below an oysterbed at Badalpur, on the coast, and at Jalpur, about 90 km inland on the river Bhadar. These were made using a light stone hammer technique and had their edges retouched by pressure flaking. They have been dated at 56,000 BP and resemble tools from the other side of the subcontinent in the Rallakalava basin near Madras and elsewhere in India. At this period of rapidly changing climates, the relation between the toolmakers, their way of life and the coast is not clear. Coastlines were on the move and the associated fauna and flora has not been preserved. Colonisation of the Indian coast by Banda probably preceded this period and, on present knowledge, the

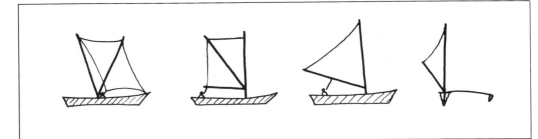

*Mid Palaeolithic tools from Jetpur and
Badalpur (northwest India) dated 56,800 BP
(a)–(d); and Middle Stone Age tools from
southeast and central Africa (After Misra
1987 and J. D. Clark 1970).*

tools imply no more than that coastal areas were a favoured habitat.

The tropical African coast is equally uninformative, because conditions are not well suited to the preservation of fossils. Further south, along the lower reaches of the Zambezi and Limpopo, the many tools and sites of the Charaman industry might be the first sign of Banda penetration into Africa.

It can hardly be expected that the Banda strandlopers would have introduced anything in the way of stone artefacts. In fact, established traditions of stonework are more likely to have been adopted or adapted

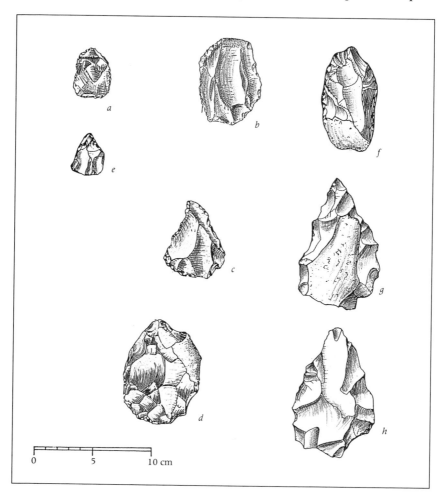

from mainland sources because seaside industries, especially those from a coral reef coast, have few or no stone tools, relying instead upon a rich choice of shells, teeth and fish bones for their hard cutting edges, and a variety of composite lash and tie techniques, using wood, fibre, cane, palm leaves and other perishables, for the bulk of their other needs.

What the Banda would have brought, however, was a new mode of subsistence and attitude to resources. While the seasides of the Cape had generated their own marginal populations of shore dwellers, the dominant economies in Africa centred on mega-hunting and for them the Banda way of life would have been new and very different. An interesting interface between the two cultures could have concerned the many uses of poison. The possibility that digging for roots was a major use for Charaman picks was mentioned in the previous chapter. If intensive exploitation and processing of roots was one of the innovations developed in Africa by Banda-influenced cultures, yam and other plant-based poisons could have been by-products quick to be adopted by the mega-hunters for whom poison was likely to have been an adjunct to hunting and trapping long before the Banda arrival. Even today various deadly techniques are employed with great insouciance in tropical Africa and elsewhere. A vivid memory of my youth is watching the Hadza hunter walking in front of me draw an arrow out of his quiver and, without breaking his pace, delicately use its poisoned tip as a backscratcher. What a symbol for the hunter's easy accommodation with life and death! That brief gesture may also illustrate what assured precision must have been applied to the smallest detail of everyday prehistoric life.

Lupemban sites are not rare and there is the strong implication that the populations that made this industry became quite dense and widespread. Relatively dense populations that are supported by a reliable and well-endowed ecosystem tend to divide up into territorial units and devise complex rules for both group and individual behaviour. Intensified social pressures and a proliferation of 'tribes' could have had their influence on tools in interesting ways.

From a continuum of tool types, such as scrapers, scrapers-cum-awls, or scraper blades, there began to be a much clearer separation of function. Tools diversified, becoming more carefully fashioned according to more or less standardised patterns, each serving a narrower function. Over quite short periods of time, and from region to region, new refinements or variations of shape were invented to replace older types. This wider variety of tools clearly performed their function more efficiently than the generally larger, multipurpose tools that were typical of the early phases of the Middle Stone Age. Nonetheless, the makers of the latter industry would have been more nomadic and were probably content with the self-sufficiency that could be achieved with a minimum of means. Thus, nomadic big-game hunters could have coexisted with more settled foragers for many thousands of years. Indeed, in southern Zaïre there is evidence of two distinct cultures coexisting in the same area but in different habitats. Here, the large two-faced stone tools of typical game-hunters have been retrieved from more open savannas. The forested valleys that traverse this raised southern margin of the Zaïre basin yield smaller, more finely wrought blades and miniature artefacts which are more appropriate to a versatile small game and plant regime.

There have been many situations in Africa, Eurasia and Australia where two sets of distinct but contemporaneous tools are found in neighbouring but ecologically distinct settings. In northern Australia, Aborigines alter their toolkits as well as their diets during seasonal movements between coastal and inland sites. Such versatility may have been common in many prehistoric people but it does not itself disprove that more ecologically specialised communities could have existed side by side, because there are ample surviving examples of separate economies running in tandem in parts of Africa and Asia.

Not all Lupemban tools were lightweight, and both hard woods and earth can be resistant to anything but the heaviest and sharpest of picks and hoes. For example, if weight is lacking, even a strong man can make little impression with a digging stick. The earliest signs of disc-shaped boulders with a hole bored through the centre have been found at Polungu in southwest Tanzania and dated to about 40,000 BP. Their function almost certainly was to give weight to digging sticks but, unlike the Charaman and Sangoan picks, these would not have been throwaway items. If their use by contemporary people is any guide, they were probably used by women, and mounted around a hardwood stake would have been more portable than cumbersome rocks. Nonetheless, they imply frequent use within a fairly small range, so that digging for roots and tubers would have been a regular activity in the areas where bored

Some Lupemban stone tools (After J.D. Clark 1970).

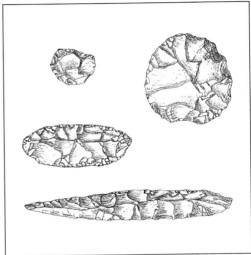

stones have been found. Extrapolating from this, it is very likely that there were belts of country in the seasonally moist tropics where wild roots and tubers offered a large supply of staple food and the opportunity to become relatively settled.

A later manifestation of the Lupemban industry in Zaïre is known as Tshitolian (dated 27,000–14,000 BP). In place of prepared rounded cores for striking points and flakes there was an increased preference for making very small, standardised microliths from pyramidal cores. The very numerous microliths known as '*petits tranchets*' were probably used as arrowheads set in mastic. Unmistakable arrowheads, complete with tangs, also appear in Tshitolian sites but these could reflect contact with earlier Saharan hunters, the Aterians.

It is generally agreed that bows and arrows have been invented more than once. Tanged flints of arrow shape and size turn up in Angola about 30,000 BP. The forest-dwelling people were evidently exploiting what is a major source of meat in many forests – monkeys. Trapping monkeys is generally ineffective because of their intelligence, sharp eyes and ability to learn from past events. Shooting them with poisoned darts or arrows has more success and the Tshitolians of equatorial west Africa made fine tanged arrows, implying that the stone technology of the earlier Saharans might have filtered south. A problem common to both

Bows and arrows of various sizes deigned for a variety of preys and ranges.

desert and forest of getting at out-of-reach prey found the same solution in bows and arrows, but the preference of some modern Pygmies for plain wooden arrows tipped with poison suggests that wholly wooden bows and arrows may be of much greater antiquity in the forest.

Stone arrowheads appear in the Cape about 9,000 BP but by that time it can be assumed that wooden arrows, albeit without stone tips, had been in wide circulation for many tens of thousands of years. Khoisan, Spanish and other regional rock paintings depict bows and arrows, and the distinction between hunting bows and fighting bows is sometimes shown.

Contemporaries of the Tshitolians in southeastern Africa developed a stone industry in which the chips are so small that they can only have been mounted in series to make saws, serrated bayonets or toothed clubs. Grinding stones for crushing cereals are also very common. This culture, which flourished over a large part of eastern, central and southeastern Africa between 20,000 and 12,000 BP is called Nachikufan. Many of the people living in these areas today must be direct descendants of the Nachikufans.

Late Stone Age artefacts that are not made of stone have had a better chance of surviving in the cool dry climates of southwestern Africa. It is here that a hint of the diversity of Stone Age life can be appreciated: mattresses of leaves and grass, vegetable debris, bark trays, pegs, wedges, mallets, leather capes and bags, bits of clothing, pigments, pestles and mortars and the oldest naturalistic rock art so far found anywhere. These paintings have been dated to about 28,000 BP and are painted on the walls of a cave named 'Apollo II' in the south Namib Desert. The artists were hunters of large game and would have occupied the area during a period when warm moist conditions were at their peak in southern Africa. The earliest trace of mastic on a stone tool comes from Apollo II but hafting grooves and tangs are apparent by 50,000 BP and the principle of tying or gluing handles to implements probably long predates this, especially in seashore and riverine industries.

The period between 75,000 and 30,000 BP saw a proliferation of regional industries all over Africa. In the forest galleries and moister woodland areas the importance of cutting and shaping wood is evident in stone axes and adzes and beautifully made, daggerlike knife blades. The implication in the structure of these tools is that they were bound

and hafted into wooden or bone handles. Hafting becomes certain with the appearance of tangs both within and south of the Sahara, but less efficient forms of hafting must have a very ancient history in the Middle Stone Age. It is a haft or handle that transforms a blade into a true knife with greater leverage and a stronger slicing action than a hand-held flake.

The Aterian industry that developed in the Sahara may have lasted a very long time if its beginnings go back some 70,000 years, as has been suggested. The preparation of tangs for attaching shafts and handles on to points was a specialisation that was consistent with greater use of projectiles in drier, more open habitats.

The Aterian had developed as a Saharan speciality. In North Africa it had replaced the much more generalised and conservative Mousterian, which may have been developed first by African Heidelbergs. The superbly made, tanged points and supposed arrowheads of the Aterians were made as light as possible and were clearly carefully tended and valuable. They were mostly made in a vitrified silica tuff, and the raw material was sometimes carried as much as 60 km to be made up. The open dry landscape of the Sahara causes animals to be more dispersed, mobile and out in the open, so that hunters too must be able to walk long distances and their exposure demands that they devise special methods to catch prey. The development of Aterian points would have been a part of this challenge.

Snares and fence corrals are likely to have been the most productive harvesting device. Natural and artificial ambushes would have assumed special importance, because of the difficulty of getting close enough to the prey. In these circumstances increasing the distance at which a weapon was effective became very important. It is unlikely that large numbers of Aterians could have been supported even in a partly wooded, temperate Sahara without their having recourse to many types of traps, projectiles and poison.

The subterfuge of dressing to imitate a bush, an antelope, ostrich, bustard or ground hornbill was widespread until quite recently. Such devices are also likely to have been a part of the Aterian repertoire, but spear-throwers or bows and arrows would have been a necessary adjunct. The finer and lighter the craftsmanship, the more likely it was that crude stabbing and slashing was replaced by accurate marksman-

Sketches of some snare and log-fall traps (top left) Log-fall from Sepik, Papua. (top right) Log-fall trap. (centre) Simple neck snare from Tanzania (from M Kakeya). (bottom) Leg snare from Kigezi, Uganda.

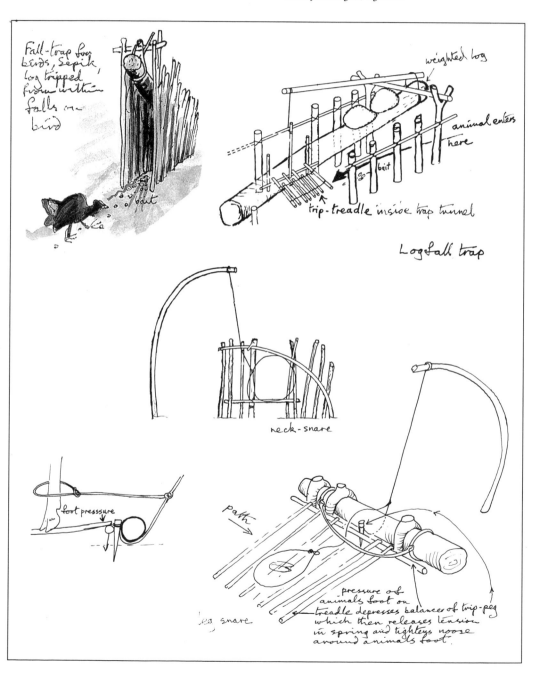

ship where the driving force was a neat application of mechanical and anatomical principles and not the exertion of muscle power alone.

Meanwhile developments had taken place in the Middle East, where tools had started to diversify about 43,000 BP. In addition to standardising a variety of chamfered blades, stones were splintered, cut, sawn, ground, grooved, perforated and polished into mostly small components, many of which may have been mounted in handles to serve as saws, knives, chisels and shavers.

It is now a familiar concept that artists and artisans can chip away at a block of inert material to 'release' a preconceived idea. The idea may be a sculptural image or it may be the ideal properties or specifications for a mechanistic tool. The latter requires a step-by-step, planned application of skill and foresight. The products of such processes are qualitatively different from rough-and-ready modifications of natural materials to assist the hands in routine performances such as crushing, cutting, levering or scooping. The latter are activities that are easily copied and practised by all members of the group, as and when circumstances demand. More complex tools for more specific tasks tend to need less easily found material than the nearest river pebbles. Relatively rare material must be collected, transported, perhaps stored, so that special

expeditions may be necessary. Economies in weight can be appreciated by both makers and users. The combination of more refined specifications, localised manufacture and specially learnt procedures makes it more necessary that the tool is made to a sort of mental template which guides the manufacture of that tool.

Paul Mellars of Cambridge University has shown that Upper Palaeolithic tools in western Eurasia were made by a set of procedures that conformed to recognisable patterns. He has suggested that development and change, more complex patterning and standardisation have their roots in social and symbolic factors. These might have helped develop a sharper awareness of what Mellars calls a 'taxonomy or mental categorisation of tool forms', in which the sex and status of users, raw material types and seasonal contexts all served to put industries into a larger social context.

The two classes of Middle and Upper Palaeolithic technology in Europe were made by two quite distinct types of people, Neanderthals and Moderns. The boundary was not a sudden transition to a higher plane of expertise; it was territorial displacement of one people and their techniques by another people in possession of a different and more developed technology. Between 43,000 and 40,000 BP the Aurignacian industry spread rapidly up into Europe, and this expansion into a cold climate involved the manufacture of clothes, dwelling structures, camp 'furniture' and new food processing and preserving techniques. Although the challenge posed by the climate and the northern fauna must have been influential in speeding up the tempo of technical innovation, the sort of changes that Mellars has analysed did not begin in Europe but may well have come from Africa, via the Middle East. Their development there had not been a sudden event and their beginnings could lie in protracted cultural changes and interactions deep within Africa.

There, the trend towards smaller tools and narrower functions seems to have accelerated along the Zambezi equatorial axis with the emergence of the Lupemban industry. It may be no coincidence that this is the zone where the Banda strandlopers could have interacted with the indigenous hunters and built up their first continental populations. The adoption of new ideas together with an expansion in the range of foods and greater skill in harvesting them followed a long period when very conservative industries had remained remarkably stable. The interaction of these two different subsistence systems in Africa may have been the initial perturbation that set all these events in train.

Since a wider spectrum of foods and greater technical versatility are likely to support larger numbers and improve the chances of survival during lean times, it can be deduced that such influences could have spread very widely throughout Africa. Towards the outer edges that influence would have been very dilute, but there is a rapid development in stone tools over this period that suggests that the cultures of equatorial and northern Africa and the Middle East were more in touch than is commonly supposed.

In this context, the archaeologist Ray Innskeep has pointed out to me an extraordinarily detailed resemblance between Lupemban and Aurignacian tools that could be significant. Both industries made very symmetrical 'keeled scrapers' that flaked a broad semicircular furrow out of an oval stone to create a sort of shallow gouge. This tool might have been easily hafted but the stone is also just the right weight and size to fit in the palm.

The relatively sudden appearance of new toolkits with wider ranges of implements implies a greater expansion than the mere possession of a larger inventory of stone tools. Many of these durable artefacts were essentially subsidiary to manufacturing processes that used wood, cane, grass, resin, bark, various fibres, skin and bone, including multimaterial structures. This increased range of paraphernalia also signifies a more intensive and multifarious use of resources which, in turn, would imply less nomadic habits, higher human densities and more effective penetration of all exploitable habitats. More complex cooperative cultures would be needed and these would have to develop a different social behaviour to cope with more frequent intercourse, more competitive disputes to be resolved, more effective and perhaps more subtle psychological relations.

The relationship between tools and human psychology has had scant mention, yet it is here that the motivation to improve tools and performance would have been most acute. It was central to the way in which children integrated themselves into the society through a proud display of skills. For example in some central African societies children regularly impose obligations on their elders (often maternal uncles) by presenting them with small game, thereby winning special approval or reciprocal favours through their skills at trapping or snaring. Adults too would have been susceptible to social incentives to outshine others. For example, one reason for improving the design or seaworthiness of a craft would be the prestige than an expedition leader gained from an important ceremonial or hunting event (such as collecting birds' or turtles' eggs during a very short harvesting season). Such expeditions would have been planned and prepared well in advance, because failure reflected shame on all members, especially the more senior. Ensuring greater stability, buoyancy or speed may be necessary for utilitarian reasons but most people will find no difficulty in identifying with the prehistoric man who did not want to make a fool of himself, sought his fellows' approval and tried accordingly to trim his boat and improve his performance.

Miniaturising and diversifying tool types would have had far-reaching social effects. The increase in skill needed to make and use Upper Stone Age artefacts would have set off strong behavioural impulses which in turn would have provoked still greater inventiveness. For example, a prime tenet in the child's upbringing would have been the public teaching and learning of a wide range of skills. Displaying a good command of these would have been more of a measure of social standing in this society than it would in groups with a more rudimentary toolkit. Public display of an increasing number of technical aptitudes would have become the vehicle for closer cooperative effort and for an overall enhancement of the quality and quantity of information being exchanged. This would have provided a big incentive towards still further innovation, even if these were merely quirks of style or nuances of expertise.

Along with the practical aspects of skill learning must have been some awareness that technical expertise bestowed power, and that this could have mixed social consequences. We cannot know whether prehistoric societies found moral dimensions in technology but a latent potential for good or ill in technology is recognised by some Zaïrois in the single word *pomoli*. The concept of *pomoli* may or may not have been widespread in the past but it certainly deserves recognition far beyond the remote forests in which it originated and has been passed on.

In children's games and in adults' teaching youngsters about animals and plants, imitation, entertainment and humorous play are inextricably interwoven. Music as part of ritual is likely to have been preceded by music as a by-product of work, growing up, learning and simple *joie de vivre.*

The first hint of a musical instrument comes from the North African coast, at Haua Ftea, where a hollow bone, pierced in the same manner as a modern flute, survives. The evolution of bone flutes could very well be a by-product of people's passion for bone marrow. In the course of sucking and blowing to get at this most exquisite of delicacies, the acoustic properties of a resonating tube would not have been missed. The approximate age of this flute is in the region of 85,000 years. Percussion must, however, have been among the earliest forms of contrived music. So many activities would have depended upon striking one material against another and many young animals, especially apes, enjoy their ability to make such sounds. Even today, it is possible to find every gradation between talking-while-you-hammer, calling-while-you-walk and rhythmic work coordinated by drumming or stamping.

Discovering acoustics through test-blowing a hollow stem to see if there's an insect, grub or mouse in it, scraping out a tortoise shell, making a rock flake ring, a hollow log or buttress boom, plopping a cupped palm in water or listening to echoes in a closed valley would have been a normal part of prehistoric experience. Imitations of bird and animal noises are also universal, and onomatopoeic names are better sung than said. Improvising imitations and entertainments might have begun as child's play and I believe that children and childhood were at the heart of prehistoric existence, its continuity and its development.

In the history of civilisation invention of the wheel has long been seen as a landmark discovery. Yet, long before the Aztecs and Columbus, Amerindian children played with toy animals fitted with wheels for legs. The baked clay artefacts excavated at Tampico, on the Carribean coast, appear to have been made by children. Even if they were cult figurines as some archaeologists have suggested, playful ingenuity is their most obvious feature. The technical potential of wheels could only be realised when there were pressing social incentives and large domestic animals to harness in front of the first cart or chariot. If the Tampico toys illustrate technical content they also show the potential for development in 'mere child's play'.

As in all prehistoric societies, the main skills required for survival in a difficult world had to be learnt within the family or clan circle. Any natural catastrophe that wiped out a particular class of an isolated community (say, all the adult men) would have been a major trauma that condemned the survivors to relearn many skills from scratch. The isolation of some Australian and Oceanic groups may have made them particularly prone to such risks.

For people closer to the mainstreams of human migration and cultural influence, a steady speeding up of the tempo of change was inevitable. The relationship between cultural influence and ecology could have been a crucial one and the ecological separation of different subsistence economies is relevant to a frequently recurring question – how exclusive was technology to a particular ethnic or regional group? It would seem that prehistoric technologies were more likely to spread faster across geographic, ecological and racial boundaries when they were directed at more effective exploitation of a common resource (such as killing large game animals). They would have been slower to cross such boundaries when innovations focused on more specialised and less universal ways of earning a living (such as techniques for climbing tall forest trees, catching marine turtles or desert rats). Distinct cultures could live their separate existences at very close quarters so long as their economies were specialised to exploit separate resources or habitats.

However, small technical advantages very quickly resulted in territorial or demographic expansion. In such situations, there were strong incentives for the neighbours of technological innovators to adopt new techniques or tools, especially if they were threatened by them and these spread very quickly.

From prehistory until the present, societies have had to keep up to date in their food-getting methods, in their modes of maintaining social cohesion, in striking alliances and in their arsenals. Any useful gadget first benefited its inventors and their immediate circle. Such benefits or advantages would have quickly distinguished their possessors from their neighbours. Where the benefit concerned getting more food and territory, allowing more children to survive, putting more people on less land or conducting a faster fight at less cost to the home side, then the neighbours either declined in numbers and influence or were eclipsed in battle.

History offers numerous instances of more efficient weapons winning decisive battles (as at Agincourt, where longbows won the day against cavalry because of their fire-power and accuracy). More recent Agincourts were the awesome 'turkey shoot' in Kuwait and the blasts of Nagasaki and Hiroshima. If the child that went fishing for tiddlers with cobwebs is brother to the man who invented the atom bomb, none of us can ever really escape the challenges of *pomoli* in a proliferating technology.

6: Is Adaptation Real?

In 1775 Immanuel Kant announced that humans and human races were the result of adaptation. He thought adaptation was real and supernatural explanations unreal.

Kant followed his contemporary Linnaeus in making four divisions of mankind but he saw each as the product of climate in a different region of the world. Superimposing a different combination of temperature and humidity upon each of the compass points, he saw the hot wet axis producing 'Indians', heat and drought 'blacks', 'whites' from wet and cold, while 'Mongols' were the outcome of selective pressures in a dry cold climate.

For a real world and real processes of adaptation it was over-tidy but Kant was challenging the religious and philosophical authorities and contradicting the conventional wisdom of his time. He was asserting that humans existed as mutable beings in the context of long past time and that there was a special relationship between habitats and what sort of people developed in them. In both respects he was right. Even if his details lacked the substance of research and intimate knowledge, it was a remarkable insight for a closeted European writing nearly a hundred years before the 'Origin of Species'. Kant failed to address ultimate origins, so what was left out of his quartet was a portrait of the conditions in which our *common* characteristics emerged. He was in no position to know that we are all Africans.

Kant was right in that human populations subjected to extremes of climate for long periods *have* developed appropriate genetic adaptations. Local populations have also developed genetic defences against other external hazards such as disease, parasites and the peculiar chemistries of a diet that is infinitely varied from place to place.

A question that highlights the nature of adaptation is to ask how an environment affects growing children. They would have been among the first to die from disease, parasites, stress, exposure and all the many vicissitudes that their parents' way of life exposed them to. Among the small proportion of children that grew up (to have, in their turn, further clutches of vulnerable children) would have been those that were in possession of some minuscule molecular advantage.

In all aspects of adaptation there are many new and developing sources of information, especially in the burgeoning sciences of physiology and molecular genetics. Blood groupings and enzymes, which are routinely monitored in hospitals all over the world, have allowed global maps to be constructed which reveal all manner of unforeseen patterns from transcontinental gradients to sharp and totally unexpected boundaries within apparently homogeneous populations. However, these patterns (which are generally based upon the *frequencies* of gene-types within a population) tend to confirm that genes vary most at the level of the individual.

Genes change rarely and at random. Furthermore the majority of mutations (especially those induced by radiation) are harmful. These tend to destroy their carriers through malfunction, in the womb, at birth or later under some form of natural challenge. Of those genetic changes that allow their carriers to survive the great majority are probably neutral. Only a tiny fraction of this fraction offers advantages but it is the carriers of these favourable or adaptive genes that are most likely to prosper and proliferate.

These mutations are the ultimate mechanisms behind the diversity of species and behind structural changes within a species. Mutations have increased the number of genetic forms, generating thereby an ever larger variety of types. Since successful mutations are so rare most of the variety we see (notably within families and tribes) is a result of recombinations of established parental genes. There are immense numbers of such gene permutations and they create unique individuals but they do not feed new genetic information into a population. By contrast migrants may transfer entirely new genes from one population to another.

Migrant genes in a new setting can be a negative liability, neutral or positively advantageous but the end product of mixing genes from different gene pools is to steadily increase diversity.

Like any other animal, a human is shaped by what he or she inherits from the immediate parents. Each of us connects, through the two parental cells that gave us life, back through more and more parents until eventually all our ancestries converge on a single pair, our symbolic Adam and Eve. It was they who ensure we would all be more alike to one *(left) Meiosis, the process of cell division whereby daughter cells have only half the chromosomes of the parent. (right) 2 examples of dominant and recessive inheritance. Carriers of one or two dominant genes (solid symbols) always show dominant trait. Recessive trait appears in offspring if both genes are recessive (open symbols).*

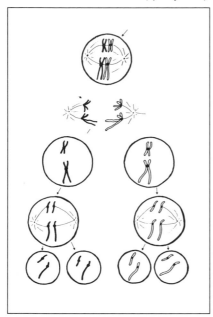
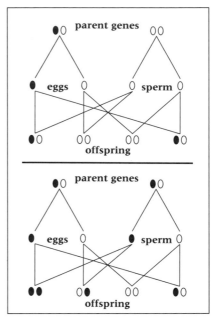

another than any one of us is to a newt, a night monkey or, for that matter, a neanderthal. If the children of Adam and Eve differ from one another today, it is partly due to recombination and partly to innumerable tiny changes in the chemistry of our genes that have taken place after our ancestors' long-forgotten copulations under an African moon.

Genes encode all the information needed for the elaboration of a foetus, the growth of a child and the proper functioning of an immensely complex organism. The genes are particles within the chemical structure of the chromosome. Humans have twenty-three pairs of chromosomes (chimps have twenty-four). These are microscopically visible messenger complexes that pass from cell to cell and are exchanged during every fertilisation.

The action of genes is essentially chemical, tissues are built by proteins made up of long strings of amino acids. The synthesis of the latter is directed by the chemical molecule known as DNA.

Judging whether a gene may have an adaptive value or not is usually a matter of inference but here is an illustration of the chemical basis for genetic change and genetic adaptation. Normal blood cells are globular.

During synthesis the exact sequence of a long chain of amino acids is vital for correct function. In some people a mutation in the genes' instruction code has inserted valine instead of glutamic acid at this critical point. The result is a distortion of the shape of the red cell globule which is called sickling. Detailed study of such conditions as sickle-cell anaemia has helped reveal a great deal about the workings of inheritance in humans.

Genes are either dominant or recessive. The simplest expression of these two modes of inheritance lies in the statistics of their outcome. The child of parents that share a trait determined by a single dominant gene has at least a 75 per cent likelihood of inheriting it. If only one parent has it, the child's chance of possessing it is 50 per cent. The percentage ratios of inheritance for recessive traits are appreciably lower and less predictable in their outcomes. Typically recessive traits are masked by dominant ones and appear only when both parents are carriers of the recessive gene. When a decisive selective advantage is associated with dominant genes and their influence is projected back into prehistory, it becomes much more comprehensible that small packets of advantageous genes should spread widely, even through entire continents.

Many characteristics that we might think of as a single entity are actually controlled by more than one gene. For example, it has been calculated that three or four genes are involved in determining human skin colour.

In any discussion of human differences it is essential to discriminate between what is changed by a mutation (regardless of whether it is advantageous or not), what is changed by natural selection and what is changed by the environment.

Take the following three sources of short stature. First there is the well-known phenomenon of human dwarfism. This condition, known as chondrodystrophy, is not advantageous and is caused by a single mutant gene that arrests growth in the long bones. Second, in several parts of the world, mostly in dense tropical forests, natural selection has favoured an overall reduction in proportions and stature. The Baka, Batwa and Bambuti pygmies of equatorial Africa exemplify these genetically determined traits. Finally a poor diet, inadequate sunlight or lack of exercise in infancy can lead to stunting. Such a failure to reach normal proportions is not genetically controlled.

Sorting out the sources of our many different traits can be perplexing but the discipline of genetics is central to this task. Mapping the human genome (a genome is the total genetic complement for a species) is now the object of an ambitious international effort. Meanwhile we can examine some grosser manifestations of the modern human genome.

Start with a mirror. That petite chin and those small white teeth are the direct bequest of ancestors who slowly but steadily emancipated jaws from their original multiple functions. Their cooking fires burning a million years ago, their knives, pounders and grinders began to relieve our built-in processors, teeth, of their heavier duties.

The decline of human jaws and teeth is part of a general refining trend in human evolution. Paradoxically, the advances in techniques and organisation that have become our central hallmark have also tended to weaken us by comparison with our immediate ancestors.

The shifting angle and declining size of our muzzle have been achieved through a neat property of bone physiology. In the course of individual development, as well as during evolution, genes control whether bone surfaces get pared away or built up. The technicalities of this remodelling are analogous to a sculptor working with gypsum plaster. Where a surface needs building up, active cells plaster it with bone salts. These cells are called osteoblasts from the Greek *blastos* meaning 'sprout' or 'bud'. By contrast osteoclasts (from the Greek *klastos*, 'to break') are erosive cells that actually nibble surfaces away. Such addition and subtraction is highly selective in where it takes place. Thus the deep-rooted teeth of apes have their bedding reinforced by osteoblasts which lay down bone all over the frontal surfaces of the face. As tooth roots declined in size, osteoclasts took over from osteoblasts to eat all that redundant bedding away. The anthropologist Tim Bromage can recognise this threshold between building and dismantling in fossil muzzles. It seems to coincide with the distinctions we make between *Australopithecus* and *Homo*, with the exception that he also found signs of local resorption above the reduced incisors of Nut-crackers (*Paranthropus*).

A shrinking muzzle has had other side effects. Where hefty teeth are deeply rooted in weighty jaws and served by densely packed muscles, the muzzle needs firm anchorage on the skull. Bony brows help to absorb and dissipate these forces. Less stress from chewing may have reduced the need for big brows, but as so often happens with redundant structures, they do not necessarily go away all at once, or evenly. The

Simple permutations of brows and teeth
(a) small brows, large teeth (narrow face);
(b) large brows, small teeth;
(c) small brows, large teeth (broad face);
(d) large brows, large teeth.

useless side toes of horses took many millions of years to disappear, so the unevenness with which big brows crop up need not surprise – like the vestigial tail that we all have to a greater or lesser degree, they are part of our evolutionary baggage.

Brow ridges sometimes correlate with generally heavy muscle and bone development elsewhere on the body. Pathological or hormonal conditions that encourage bone deposition are known to pile on growth above the orbits. Powerfully built males are more prone to develop brow ridges than women and this tendency can be individual, influenced by genes, by hormones, occur regionally or be a combination of all these factors. Furthermore, heavy build may have become more widespread during prolonged periods of peak nutrition and declined during less favoured times, so the size of brow ridges in fossil skulls may have waxed and waned in the past.

Periodic fluctuations in body size could have had interesting side effects. Enlarging or diminishing total size or weight seldom involves an even increase or shrinkage of proportions. The resulting changes, known as allometry, can be very slight or quite extreme. For example, in different populations the relative size of teeth or jaws could have altered as their overall mean body size fluctuated over time. Meanwhile another feature, also linked with body size, might fluctuate independently. This disjunction between two characteristics can be illustrated with brows and teeth. Africans tend towards small brows and big teeth; many Europeans combine either large or small brows with small teeth; Mongols have a small-brow-large-teeth tendency and Australian Aborigines tend towards a large-brow-large-teeth combination. Thus every

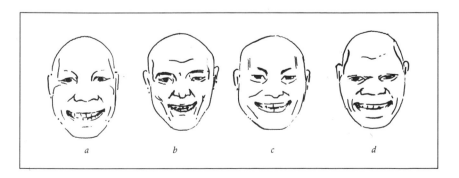

a   b   c   d

permutation of these two features occurs. Both are in an overall regression in the human race but both are subject to independent variation. Extremes of size and body build may have coexisted at quite close quarters in the past (as indeed they do today along sharp ecological boundaries in equatorial central Africa).

Short, lightly built people have many advantages; they can move quickly, climb more easily and survive on smaller rations. In some climates they suffer disadvantages; they get chilled more quickly in extreme cold and lose out in any direct confrontation in which physical strength is tested and its outcome is decisive for control over resources. The name 'Pygmy' has been conferred on many people who grow to no more than about 1.5 metres (about four foot six) in height. Although most pygmy characteristics are genetically controlled and the result of natural selection for smallness, there are also factors in their environment that favour limited growth or stunting.

There seems to be a relationship between people of shorter stature and high-rainfall forest, especially on mountains. Heavy rainfall leaches out many trace elements and minerals. The combination of a dense canopy and frequent cloud cover reduces exposure to sunshine with further implications for vitamin D metabolism. Adaptation to such difficult conditions is likely to take a long time and it is probable that all pygmoid populations are of relatively long establishment. Dwarfed populations have arisen independently in mountain and lowland forests in Africa, in the Andamans, Malaya, the Philippines, Papua New Guinea, the Solomon Islands, the New Hebrides and northern Queensland. Lack of certain nutrients and direct sunlight seems to be one common factor but even here there are exceptions (such as the well-fed shore-dwelling Andaman Islanders). In most cases the genes of these people now have many resemblances to those of their larger neighbours but nonetheless there is little doubt that they are mostly relictual cultures.

Extremes in body build often combine with extremes in climate and the contrasts sometimes make good sense in terms of physiology. Short, heavily built bodies with the ability to lay down fat have proven survival value for the Arctic Eskimos, because such bodies conserve heat well. By contrast, long, thin, athletic bodies lose heat rapidly. The Nilotic people, who represent an extreme in these proportions, live around rivers and lakes in equatorial Africa. Originally hunters and fisherfolk in very exposed surroundings, they lived off mobile and unpredictable resources that would have demanded frequent and lengthy treks as the herds migrated and the floods spread or receded.

Let us return to the human face, wherein we seek so many clues to identity, rapport or hostility. A face carries enormous emotional impact – people respond more to nuances of expression in a tiny area of the face than they do to much larger cues. It was a favourite device of early cinema to film two actors from a distance, one calm and unconcerned ignoring the other's wild gestures. A close-up shot then focused on the quiet one, who floored the other with some subtle eye, mouth or finger gesture.

Eyes, mouths and their immediate surroundings are the only organs of expression – the rest is gesture, body language. Yet the detailed structure of eyes and mouth is not very different between human and ape. The primary functions of seeing and enclosing the teeth are far too important to stand much modification for a secondary function such as communication. Besides, the communicative powers of eyes and mouth are already highly developed in apes and many other animals (in fact, a lot more subtle in a gorilla than in the average comedian). Dominance and submission are very readily conveyed through eye contact, no less between humans than animals, and it is in the continued usefulness of hierarchy and dominance relations that eyes and eyebrows play a role. The Soviet president Leonid Brezhnev was quoted in the popular press as boasting that he got to the top of the pack thanks to his eyebrows.

How could facial expression have played a part in human development, and do the differentiation and variation we see in human features have any adaptive value?

Begin with the eyes. The normal human eye, like that of most mammals, is brown or near black and this pigmentation with melanin is the equivalent of built-in dark glasses. Most mammals with pale-coloured eyes, such as cats, are nocturnal, so the physiological usefulness of brown eyes in sunny habitats is not disputed. The majority of birds also have brown eyes but there are a large number of species that have brightly coloured irises, from white and yellow to green, blue, purple and orange. In some species of storks the sexes are identical except in the colour of the eyes; in other birds eyes alter their colour with changes of mood, age or status. Virtually all birds with coloured eyes appear to use them during courtship, territorial display or to help regulate status in the flock.

Human blue eyes now occur very widely in the world and the mutation that depigments eyes and makes them blue has almost certainly arisen more than once. A localised genetic mutation is thought to be the explanation for isolated pockets of light-coloured eyes in Africa, India and among some northern Amerindians. Assuming that infusions of external genes can be ruled out, it is obvious that these mutations have failed to become dominant in the way they did in northern Europe.

It is more than likely that the vast majority of blue-eyed people owe them to a very few founder ancestors, possibly just one, in an originally isolated population. Just as Australian Aborigines owe their genes to a handful of early immigrants, so these depigmented eyes first appeared in a pioneering community at the northern frontier of human expansion in Europe. This genetic mutation must have occurred after 35,000 BP and probably after the deglaciation of the Baltic Sea about 16,000 years ago. It then spread along the shores of the sea, where, even today, more than 80 per cent of the population still have blue eyes.

Since it is certain that brown eyes are a better all-round protection, especially in bright sunlight, why should blue eyes be so prevalent in just one part of the world? Physiological and environmental explanations for blue eyes have been attempted, but the best that can be said of the Scandinavian climate is that natural selection against non-brown eyes would have been weak, because summer is short. If a single mutation is explanation enough, then the blue probably appeared first as a nonfunctional by-product of another more explicable genetic change – overall depigmentation. If a neutral mutation is unsatisfactory, return then to the expressive potential of eyes. Here humans can again be compared with birds.

Consider the very similar brown and black sicklebills of the New Guinea mountains. These birds of paradise separate ecologically on an altitudinal contour just above 2,000 m, the brown living above, the black below. The most striking difference between these two birds is not the somewhat lighter colouring of the former but the brown's brilliant blue eyes and the black's vermilion ones. Given that the colours of male birds of paradise are a prime illustration of sexual selection, the development of blue and red eyes in males can be confidently ascribed to females making different choices at the two altitudes. What is interesting is that the blue eye colour should be linked with paler plumage, because human blue eyes are decisively linked with diminished pigmentation in the skin and hair. This suite of three albinoid characteristics has much the same frequency and distribution around the Baltic Sea. As people moved north, paler skins would have been favoured by the body's failure to synthesise enough vitamin D. Since the chemical reaction that produces this essential vitamin depends upon sunshine falling on the skin, masking pigment became a disadvantage. Paling of the skin was therefore functional, while blue eyes and fair hair were more likely to have been adjuncts, genetic by-products of this necessary paling which only later became emphasised and selected through sexual preferences.

The original founding population in which these characteristics arose must have been very small but placed in a strategic position to colonise new land as southern populations moved north. The convention of choosing blonde mates must have been very strong among these prehistoric hunters because a contemporary frequency of 80 per cent implies virtually 100 per cent before more recent centuries of immigration and genetic mixing had their diluting effect.

Quite apart from the fact that blue is a bright 'advertising' colour, paling of the iris changes the main boundary of the eye from iris to pupil. This is a significant change, because the iris is of fixed size, whereas the pupil enlarges and contracts under two influences – light and emotion. Pupils send better messages than irises, especially aggressive ones, and the phrase 'he fixed him with a gimlet stare' describes a smaller black point encircled by a still, pale glare.

Sexual selection can generate considerable divergence among isolated populations, such as those on islands, peninsulas or in ecological enclaves. The Scandinavian example suggests that so-called 'racial' types that now inhabit large areas of the modern world could owe their distinctive appearance to quite arbitrary aesthetic tastes in small and originally isolated prehistoric tribes.

Value judgements on what is 'beautiful', 'impressive', 'correct' or generally 'good' are often very localised, and face-type preferences are betrayed by traditional words, songs and metaphors that extol the desirability of males or females. A pale, 'moonlike' face with no prominences, dark eyes and long, straight, black hair are the choice in

Map of northern Europe showing percentages of population with light eyes (blue) (in part after Coon, Garn and Birdsell 1950).

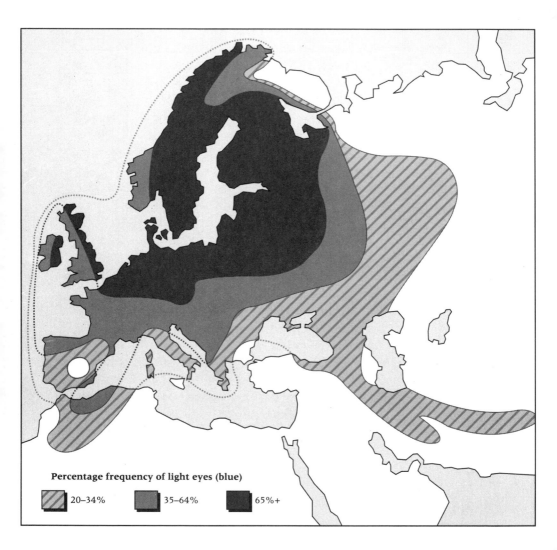

Percentage frequency of light eyes (blue)

20–34%   35–64%   65%+

China; 'gazelle' eyes, a bold nose and an angular wedge-shaped head are admired in the Caucasus; skin 'like an aubergine' and a round face with splayed teeth form the epitome of beauty in the Nile Sudd. There is no shortage of illustrations to confirm Darwin's suggestion that local aesthetics could have been a selective force in human evolution.

For those who would seek too close a relationship between facial appearance and the environment, the contrasts between northern

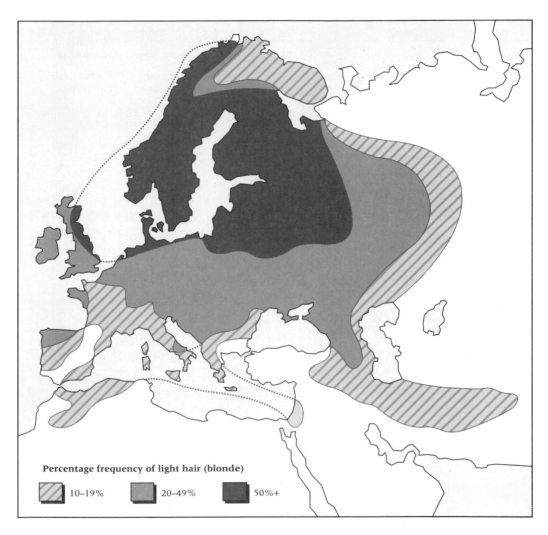

Map of northern Europe showing percentage of population with light hair (blond) (in part after Coon, Garn and Birdsell 1950).

Percentage frequency of light hair (blonde)

| 10–19% | | 20–49% | | 50%+ |

Europeans and northern Chinese are a warning. The former show a mix of characteristics with some, such as blue eyes and blondness, clearly recent acquisitions, while others, such as hairiness and bony brows, are apparently 'primitive' traits. East Asians, instead, have a suite that includes raven black hair and dark eyes and retains quite a different set of apparently 'older' features, such as broad, flat cheeks and shovel teeth. The argument that hairy people keep warmer in the cold has no

231

substance and Europeans are actually unexceptional in their adaptation to cold; long noses are prone to frostbite and pale eyes to snow blindness. Both the Europeans and the Chinese have depended on furs, fires and a good diet to keep them warm. Technology not nature was their main shield. Notwithstanding this, Chinese adaptations to cold may have a more convincing case.

Kant's idea that the Chinese had adapted to a cold, dry climate was taken up in 1950 by the anthropologists Coon and Birdsell, who argued that the rounded fatty surfaces and hairless skin of Mongols are less susceptible to frostbite, while the danger of snow blindness is also reduced by narrow fatty eyelids and very black irises. Laboratory experiments on how European and Japanese faces respond to very cold air have shown that the latter are able to maintain higher skin temperatures. This is mainly due to a better local circulation but is also likely to be part of an overall temperature response that is genetically controlled.

Originally, the earliest tropical populations of Modern humans seldom faced any risk from extreme cold. The most energy-saving response to mild cold, and undoubtedly the original human condition, is to shut off peripheral circulation and let skin temperatures drop at night. In this way Africans and Australian Aborigines can comfortably fall asleep in the open, even on a chilly night.

By contrast, those populations that moved north had to develop responses that would protect tissues from being frozen. Most Mongols, and especially the Eskimo, reduce the risk of frostbite by burning more calories and keeping the peripheral circulation going all night. Europeans also tend to maintain a higher temperature throughout the body and especially the hands and feet.

The disadvantages of the original temperature regulation system become apparent only under very cold conditions. During winter exercises the US military found that Afro-American soldiers were more prone to frostbite than those of Oriental or European extraction. The same army has found this genetic pattern reversed where overheating was involved. White Americans are more prone to heatstroke than Afro-Americans, because the latter are better at maintaining a stable internal temperature in spite of both sweating at much the same rate.

Few would dispute the proposition that the most northerly Mongols, the Inuit or Eskimo, must be the best adapted of all Modern humans to life beyond the Arctic Circle. There is an implication here that the immediate ancestors of the Eskimo and other Mongols may have had a longer period to adapt to extreme cold, but this can only be a guess while their origins remain unknown. The predecessors of the Europeans, instead, started moving north no earlier than 43,000 years ago. Since many southern Europeans are hairier than the northerners, their hairiness cannot be invoked as an adaptation to cold.

Hair is one of the most inescapably mammalian attributes. Its uneven growth on the face follows a similar pattern on some of the great apes and a tendency to frame the male face is shared with some big cats and certain monkeys. We resemble countless small mammals in having hair tufts placed above warm glandular areas to allow hormonal and other scents to evaporate and thus disperse their intimate messages.

Normal unspecialised mammal hair is designed to insulate against moderate fluctuations of cold and heat, and the helmet of hair that most people possess undoubtedly serves (within narrow limits) to keep the brain at an even temperature; significantly, both men and women have head hair. Beards are a different case; insulation and scent dispersal are out. Beards are a visual device to enlarge the face. Whether this is primarily designed to impress rival men or women is a moot point (probably both). What is interesting is the uneven distribution of beards through hot and cold climates, and among differently pigmented people. This suggests once again that selection in favour of male beards was mainly the result of prehistoric cultural preferences. Sophisticated cultures seeking a sense of order and conformity in crisp, trim images are embarrassed by this reminder of prehistoric unruliness; the answer is to shave. Cultures that seek to appear civilised but also to maintain a firm distinction between men and women compromise – the men grow moustaches.

The presence or absence of heavy beards could conceivably have a broader bearing on communication. The mouth, together with the eyes, is the main transmitter of messages. Since the mouth articulates speech it ought to be watched very closely; it conveys most of the visual accompaniment to all the talking, singing, miming and shouting that together make up speech and rhetoric. What then of the bearded man whose every expression is muffled by the overgrowth of hair? Are there any long-term or prehistoric implications in this? Will we find a reflection in the language and style of delivery of peoples with heavily bearded ancestors?

People in whom neither sex has much facial hair have more cues to follow, and the mouth, as a prime focus for the audience, can augment messages with poutings, laughs, grimaces and a play of facial drama. There is more demarcation of 'lip boundaries' around the mouths of non-bearded people and it seems possible that there was once some reflection of this in more elaborate languages and speech styles. Even today gestures and language are noticeably more dramatic in a number of tropical languages. By contrast some northern European tongues tend to lack obvious emphasis or drama. Make what you will of this comparison, but when I was foolish enough to posit that 'beard muffling' might sometimes favour the content of a message over its style the anthropologist Alison Jolly had this to say: 'No one pays attention to the content of bearded men's speech. We are too fascinated by the beard and its visual message, so the audience should agree with whatever is said without asking questions. Beardless people on the other hand say things with their mouths that should be paid attention to. Everyone (including girls and women) knows that girls and women are more verbal.' I could only back off from my idea, mumbling, 'Tom cat? Or monkey?' into my beard.

There is one facial organ, the most prominent and central of all, that has not yet had a mention – the nose. The astonishing variety of shapes and sizes found in the human nose has never had a wholly satisfactory explanation. Nonetheless, as long ago as 1923 the physical anthropologists Thomson and Buxton established that there was an association between climates and the shape of human noses. They demonstrated that both temperature and humidity correlated with their measurements of the nose, their 'nasal index'. They suggested that minor modelling of the shape of noses, both external and internal, could improve their function as air-warmers and humidifiers, important functions in very cold or very dry habitats. It could therefore be predicted that 'specialised noses' would be commonest in these habitats. A long, narrow aperture should, in theory, help to moisten inspired air (and also to warm it) better than the wide, short chamber which must have been the original tropical form. Another possibility is that narrowing of the nostrils was the adjunct of a shrinkage in the jaw region below. In any event, narrow projecting noses with small nostrils were emphatically secondary developments in Modern humans. The independence of nose form from other cranial features is also an illustration of how many separate variables there are in that mutable complex, the human face.

The face, for all its autonomy as an organ, is but a small part of the body with which it shares its most visible characteristic, skin colour. The functional explanation for humans having different skin colours is that they have evolved in direct response to sunlight exposure or sun starvation. Both extremes can carry severe costs in an inappropriate climate and the most versatile and most widely distributed skin type is a light brown that has the capacity to tan or to lighten in response to varying sunlight levels.

Among many older theories about evolutionary sequences, one was that Modern humans were black-skinned to begin with, then some lightened to brown and finally others lost even more pigment. Another assumed that skin colours were relatively rapidly selected for and that various skin pigments had evolved a number of times in various parts of the world. On present evidence Modern humans are likely to have begun with all the built-in advantages of a versatile light brown skin and only developed later the extremes of densely shielded (black) or totally depigmented skins.

It *is* a radical change for an animal to become depigmented. The fact that that type has succeeded and become numerous implies that the loss of pigment is linked with some real advantages. Once again the explanation lies in a physiological modification to offset a specific problem posed by climate.

Calcium metabolism was a special problem for people in the far north, as has been shown in numerous human remains from 15,000–10,000 BP, notably in Sweden. The bones and teeth of these prehistoric populations show signs of inadequate deposition of calcium. Too little calciferol is the most likely immediate explanation, too little sun and a poor diet the ultimate causes.

The body's control of calcium metabolism or bone-building, normal growth and kidney function are all controlled by calciferol or vitamin D3. Because it is formed in the skin through the sun's rays acting on steroids *below* the filter layers of the skin, any screening by the sun may stunt or impair growth, with serious implications for children and pregnant women. Any failure or shortfall in the metabolism of calcium during a female's life is likely to affect both herself and her offspring and greatly influence their long-term survival. While pregnant and in milk, she may drain her own resources in order to build her baby's body and teeth. In severe cases neither will get enough calcium. Even greater risks attend women with rickets. The birth canal of a rickety girl's pelvis may have no more than half the area of that in a healthy woman. In the far north the risk of death during childbirth would have been very high for prehistoric peoples and the danger from rickets could have been a major selective force favouring the development of a pale skin in northern climates.

It is possible that there were other selective advantages in depigmentation but they would have been subsidiary to calcium metabolism. Melanin-producing cells derive from the same embryonic layer as nerve cells and melanin arises from the same biochemical pathways as the neurotransmitters in the nervous system. They are both by-products of DOPA, an amino acid that is synthesised from the precursor tyrosine. It has been argued (though not so far as I know in relation to humans) that animals that lack melanin are less susceptible to stress in general and less sensitive to environmental stimuli (hence white laboratory rats and mice). In prescientific vocabulary they are more phlegmatic. Although there are many folk beliefs that describe pale northerners as less excitable than their darker cousins in the south, it is very doubtful whether such differences (if they exist) could be sufficiently marked in humans to be measurable.

Nonetheless, efforts continue to find physiological or psychological characteristics that correlate with simple visual differences. Depigmented eyes have been a favourite. American psychologists, spurred perhaps by the national obsession with competitive games, have investigated every possible clue to good form in the sports arena. One of them has claimed that there is a correlation between brown-eyed competitors doing best in fast, quick-reflex sports, while blue-eyed persons are more effective where delayed and planned actions are called for. This result is as likely to have been influenced by cultural differences within American society as by the subjects' physiology or anatomy.

Returning to prehistory and the relationship between sunshine and pigment, consider how the earliest Modern humans in Africa might have coped with sun.

The great majority of mammals shelter from the heat of the day, and humans in all the wooded mainland habitats of Africa and Asia would have been no different in ceasing activity. There would have been little incentive for a 'mad-dogs-and-Englishmen' approach to subsistence. Going out in the midday sun is a much larger evolutionary step than is generally appreciated and one that has required adaptation in all tropical animals that have attempted to maintain activity in the open at midsummer temperatures. Even oryx, gazelles and camels prefer to rest at this time. Several factors are involved; one is overheating, another is desiccation and a third is damage to tissues within and below the skin by solar radiation. So long as animals or people achieve an adequate living without excessive exposure to sun and heat, there is likely to be active selection against such aberrant behaviour, because sun-inflicted damage would normally reduce their fitness.

The 'sun taboo' was broken in Southeast Asia for several reasons. Collecting shellfish and other mud-flat or tide-pool organisms may have begun as an extra to riverine or forest-edge subsistence but it would have assumed greater and greater importance as people discovered the very large volumes of food that could be recovered. With the use of canoes or rafts this mode of existence got launched and a decisively seashore life became established all through the island archipelago. Islands then became significant at three closely linked levels: the technological, the ecological and the genetic. The technology that took people to the islands developed around ecological resources that then became the incentive to reach the islands. Cultural conservatism would have kept most of the new islanders as shore dwellers, because the techniques and cultural traditions that took them there in the first place were easily maintained and would have continued to be super-productive.

As more and more islands were reached, the populations that developed there differed fundamentally from any previous ones based on land. Further contact between island and island or island and mainland depended upon watercraft. These not only broke physical links with the mainland, they were the artefact, the one and only technical device that would then serve to maintain the genetic distinctness of islanders. The invention of watercraft introduced an entirely new contact/isolation mechanism into the course of human evolution and adaptive radiation. Genetic change was now directly dependent upon a single technological invention.

Subsistence in Southeast Asia demanded exposure to sun. In place of a day-night cycle the island shore-dwellers lived by a tide cycle and the time for foraging was not a matter of human choice or convenience but a waiting upon the tide, 'which waits for no man'. The whole community, but especially women and children, the main collectors of shellfish, would have had to go out under an equatorial sun that not only irradiated their naked backs but reflected off the sand, mud and water to add still further to the heat, glare, sunburn and desiccation.

Recent research has shown that most non-black infants have too little pigmentation at birth to protect themselves and that they can begin to burn within a few minutes of exposure. Furthermore, overexposure during infancy may be a major trigger for the deadly cancer melanoma, which develops somewhat later but often kills sufferers before they are of reproductive age.

Whereas selection had probably worked *against* exposure among mainlanders, now there would have been a total reversion and powerful selection *in favour* of sun and heat tolerance among the islanders. The importance of this reversal in selection pressure cannot be overemphasised.

Charles Darwin was one of the first scientists to suggest that dark skin protected the body from the harmful effects of sunlight. Black skins have also been assumed to be a direct adaptation to tropical climates. In 1956, J. B. S. Haldane thought such adaptations were a matter of time and he cited the tropical Amerindians to illustrate that more than 10,000 years was needed to evolve such traits. If a recent dating from a South American archaeological site turns out to be correct, this estimate would have to be tripled. In fact, the brown or reddish Amerindians serve to confirm that pressure to become black is *not* merely a matter of living in the tropics. A map of ultraviolet radiation levels over the Old World shows that there is no correlation between black skins and sunlight. Behaviour can be modified to reduce exposure and, indeed, this is what most non-black people in the tropics do. Even the Khoisan, archetypal African people, make a point of avoiding excessive exposure to the summer sun by taking shelter for the heat of the day.

So what are the stresses incurred by too much sun? Ultraviolet rays in sunshine burn naked skin and sunburn has many effects on skin. Apart from destroying cells, it can so disrupt the surface blood vessels that they become congested (a condition called erythema) or cause blisters of fluid, oedema. Elastic fibres and collagen in the skin are broken down, causing the skin to lose tone and get wrinkled. Sunburn sores can let in all manner of secondary infections (making it more difficult to offset overheating by efficient sweating). Heat exhaustion can stress the heart.

Whereas northerners get too little sun, too much can overstimulate the production of vitamin D to toxic levels. An unpigmented human body can synthesise up to 120,000 units of calciferol in one hour of total exposure to sun. This is three times the daily need, and sustained overproduction of calciferol causes various unwanted calcium deposits including kidney stones. (Calcium stones are also aggravated by too little metabolic water, another by-product of frequent overheating.) Too much ultraviolet light is also thought to destroy some essential nutrients and vitamins that are light-sensitive, notably riboflavin, vitamin E and folic acid. Folate deficiency, which is known to be influenced by overexposure to sunlight, has severe effects on fertility. Thus a folate-deficient population would decline faster than a healthier one.

By far the most serious consequence of burning by ultraviolet light is skin cancer, an uncontrolled growth of abnormal cells, which takes several forms. The commonest type affects basal cells. Less frequent are squamous cell carcinomas, while melanoma tends to begin as a dark skin freckle which later spreads into deeper tissues where it can cause death. The milder forms are often very slow to kill but, nonetheless, greatly reduce the fitness of individuals or populations suffering from them. In a prehistoric setting even the mildest of skin cancers would open the body to innumerable infections. White Americans, who are particularly given to sun-worship, have seven to eight times as many skin cancers as black Americans living in the same cities.

Melanoma cancer is more specifically tropical, develops fast and kills a high proportion of children. Until recently, more than half the whites that developed melanomas in Australia died. Even now, with the mortality rate reduced to about 15 per cent, many thousands of people have died of melanoma in Australia over the last decade. Very low levels of skin cancer among Australian blacks and high prevalence among whites suggests that natural selection is being held off by a combination of technology and medicine, but only just; in the city of Cairns in northern Queensland, one in ten 60–69-year-old white men have skin

Ultra violet radiation over the Old World. This distribution shows that black skin is not a direct adaptation to ambient levels of ultra violet radiation (which is particularly high over cloudless mountains where there is a thinning of the ozone layer).

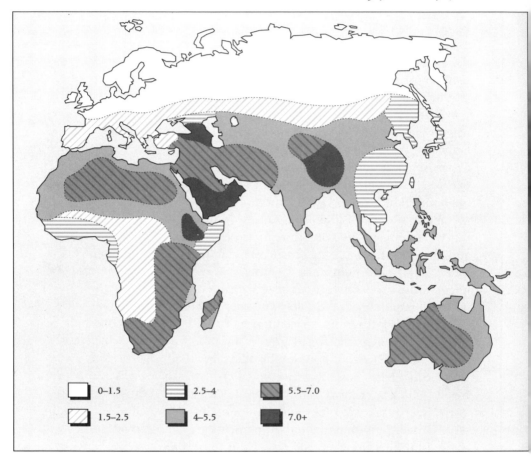

0–1.5	2.5–4	5.5–7.0	
1.5–2.5	4–5.5	7.0+	

cancer and deaths from melanoma are the highest anywhere.

In the contemporary world, the workings of natural selection on humans are masked by innumerable cultural and technological shields. Cushioned from direct experience of these lethal natural forces, some armchair theorists even deny the existence of natural selection. Yet there are many parts of the world where fragile human bodies can survive only with the help of special physiological adaptations or elaborate medical and technological aids. Barrier creams, sunshades, shirts and Ozzie hats are a cumbersome alternative to black skin and eyes and dense curly hair.

As originally tropical animals, all humans are well adapted to tolerate heat. We all share the same number of sweat glands (about 5,000). We

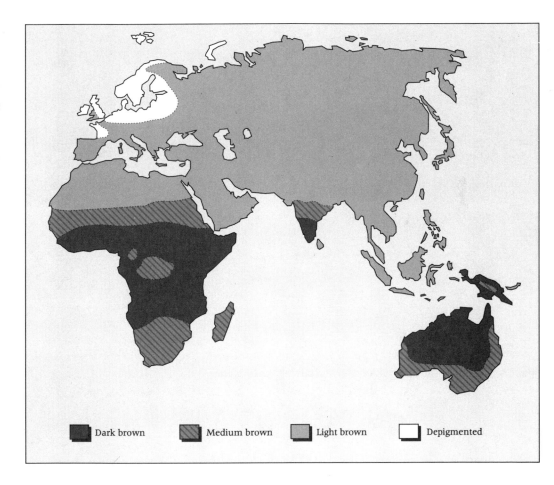

Dark brown   Medium brown   Light brown   Depigmented

also share in the number of melanocytes, live cells embedded in the deeper layer of the epidermis, where they divide and grow throughout life and produce melanin, slowly in light skins and rapidly in dark ones. Melanin has the property of absorbing and reflecting ultraviolet light, so it clearly functions as a sun filter. Furthermore, melanocytes provide the mechanism for a quick and flexible response. Too much sun and they churn out melanin; too little, and they become quiescent.

Differences in skin colour arise from different genetic patterns in the way melanin gets packaged and distributed in the lower levels of the epidermis. The pattern found in all light brown to pale skin types is that the melanocytes synthesise melanin and package it into a recognisable

The skin and ultra violet radiation. Both epidermis and pigmented layer may be thick or thin. Some ultraviolet rays may fail to penetrate to the pigmented layer.

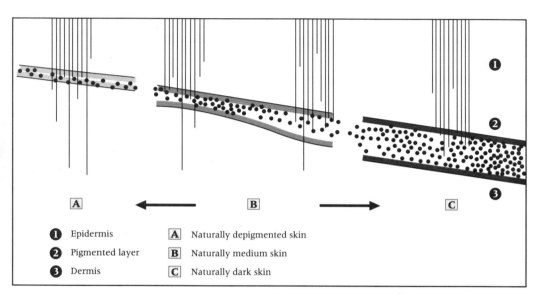

❶ Epidermis   **A**   Naturally depigmented skin

❷ Pigmented layer   **B**   Naturally medium skin

❸ Dermis   **C**   Naturally dark skin

unit (called melanosome) which then gets embedded in the lower epidermis. During this transfer the melanosomes form clusters or complexes, which are contained within a membrane.

In the course of adapting to too little light, northern Europeans reduced the size and number of melanosomes and also the overall thickness of both the epidermis and its pigmented layer. This thinning allowed more light to radiate the outer layers of the body, but this adaptation was useful only where the sun remained relatively weak and seasonal. This northwestern modification compromised versatility and the great majority of people retained a more flexible type of melanin physiology.

The adoption of a beachcombing existence with exposure to extreme solar radiation *combined* with physical isolation would have broken with all precedents. The outcome was very severe selection for a type of sunscreen that was total and permanent instead of partial and temporary.

Recent studies on the microstructure and detailed physiology of melanin have revealed that black skin represents an evolutionary innovation from the clustered melanosome complexes that typify both brown and paler human skins. This innovation can be pictured as a sort of genetic puncturing of the bags that contained the melanosomes, releasing them to form a dense and evenly dispersed mat in the epidermis. These freewheeling melanosomes differ from the previous type in the histology of their melanin: they are more numerous and larger and they are embedded in the epidermis in their own way. This type of melanin physiology is recognisable before birth and lasts through-out life. The genetics and mode of inheritance for this type of skin are still not fully understood but, by matching real results with theoretical projections (of how succeeding generations would be coloured after black and white American marriages), W. Bodmer and L. Cavalli-Sforza (1976) concluded that three or four separate genes controlled skin pigmentation, so very few chromosomal loci are involved. This is consistent with a relatively small evolutionary change in the way melanosomes are distributed being enough to turn pale-brown skins densely black. These findings tend to confirm that black skins are the more derived type.

It is of special significance that freewheeling melanosomes occur only in some Melanesians (such as the Solomon Islanders), Australian Abo-rigines, Africans and many New Worlders of African descent. The recent discovery of this hidden link between now very distant peoples is consistent with all of them deriving from common Banda ancestors. Their relationship, which can be supported on other grounds as well, makes it more likely that this was a specific and structured cluster of mutations in a single early population rather than a staggered assortment of independent mutations which have subsequently converged in differ-ent parts of the tropics.

Just as the ultra-pale skins of northern Europeans are often linked with blue eyes and pale hair, black skin is linked with very dark eyes and normally with black hair. Tests on very dark eyes have shown that they retain good acuity in the very brightest of lights. While a person of mixed ancestry can have blue eyes and a very dark skin, this combination is much rarer than dark eyes and a pale skin. It is normal for both dark eyes and a brown skin to be genetically dominant.

People with dark skins usually have black hair but there are two highly localised areas, one in southwest Australia, the other an equally isolated zone in Melanesia, where nearly all children and some adult women have pale straw-coloured hair. Half a world away from Scandinavia, two marginal populations have mimicked a feature that might have been favoured as a sign of local identity in all three areas. Unlike total albinism, the *'casque d'or'* is seen as an attractive feature in children and women, so social and sexual selection may have been quite positive.

A lack of body hair is most conspicuous in the tropics. This 'loss' is really a miniaturisation of hairs until they can scarcely be seen. It is as if baby down was stopped from growing into an adult pelt. The fact that this trait has been taken furthest in the most equatorial of people tends to confirm the link between human nakedness and keeping cool. A lot of hair can not only reduce the cooling effects of sweating but also provides a purchase for skin infections via the follicles. When a body is subjected to a temperature of 38°C it can push through as much as 28 litres in twenty-four hours merely by sweating. That sweat carries away as much as 15 per cent of the body's total salt reserves as well as unwanted waste products. It is clearly preferable that such salts and wastes should wash off rather than build up in a mat of hair.

The one area where hair is really important in a hot climate is on the head and blacks, in spite of being the most hairless of peoples, have the densest and most matted of hair on their heads. The more circular a single hair shaft is, the straighter it will be: this is the normal hair section in many mammal species, including apes and humans. Straight hair is the primitive type and it is clearly a specialisation to flatten the shaft so that each hair grows in a tight spiral. A mat of spiral hair creates a thicker shield over the brain, thus insulating it from both heat and solar radiation.

Both these dangers were most acute for early Banda strandlopers, yet it seems to have developed later than skin blackening, because woolly hair is not an attribute of most Australian Aborigines. Where it does occur (in north Australia) it is plausibly the result of later genetic influences from Melanesia. While its limited distribution certainly suggests development after the first colonisation of Australia, it could also imply origination in a more westerly part of the Banda region. Although hair flattening is a very minor modification, its close association with a broad suite of other Banda characteristics makes a single origin in the Banda more plausible.

Nonetheless, flattening of the hair shaft is not restricted to dark hair and the probability that spiral hair has evolved more than once is confirmed by a small pedigree of spiral hair that occurs in some isolated Norwegian families. This type of blond spiral hair is inherited through a single dominant gene that in the Norwegian case must have been the result of a very recent mutation.

If shaft-flattening in black hair was selected for under extremes of radiation and heat, its place of origin is most likely in a well-islanded, wholly equatorial part of Banda. The coasts of northern Sumatra and the Malay Peninsula combine a multiplicity of islands with a temperature that stays over 26°C for both the northern and southern summers. Small enclaves of dark-skinned, woolly-haired people still survive in the vicinity, notably the Semang and Batek in Malaysia and the Andamanese. The limited survival of such people in this region could be held to contradict its being a centre of origin, particularly since flat-shafted hair is genetically dominant. In most cultures that have permitted mixing, spiral hair has spread widely. On the other hand, in societies with strong taboos on marrying outside the group, this easily recognised hair type could have combined with skin colour to demarcate a social boundary that was rarely crossed.

The rate at which populations reproduce is a compound of many factors, some of which are cultural, but at least one aspect of reproduction has a genetic and adaptive cause – a tendency towards twinning or multiple births. A hunting and gathering economy gains very few advantages through rapid population growth. Indeed many recent foragers actively controlled the numbers and frequency of children, because mothers not only have to bear and raise their children but also collect and process a larger quantity of food. Although the older children may be an asset, the child that has to be carried or otherwise limits its mother's activity is a burden. In some societies twins are therefore regarded as a disaster or bad omen that sometimes required the elimination of one or both twins.

Some agricultural peoples, on the other hand, gain extra helpers through rapid reproduction, and a man in a polygamous society can claim power and land through the number of workers he can put into the fields and the harvest they raise. In such societies multiple births could be regarded as a blessing. It is interesting therefore that there are wide differences in twinning from two eggs (twin frequencies derived from splitting a single egg are virtually identical worldwide). In China, 2.5

mothers in every 1,000 births produce twins. In Nigeria approximately 40 out of every 1,000 births are twins. Such an enormous difference implies that selection in favour of twinning may have developed since agriculture took over in western Africa, whereas the Chinese (under the influence of various cultural constraints perhaps and in spite of also being agricultural) have retained the older pattern.

If the human body's defences against climate have generated simple and obvious alterations in skin colour, hair form and face shape, the same can seldom be said for defences against disease. Not only are our genetic defences against disease hidden but so are our inherited susceptibilities.

It is a perennial miracle that the complex systems that build, sustain and repair a single human being can continue to operate efficiently throughout a long life. Nonetheless, there are genes that account for metabolic faults, and a recent count lists 2,336 conditions, most of them diseases, that can be traced to a single gene or gene pair. All peoples have a share of these; for example, genetically determined mental handicaps are widely distributed and account for 0.08 per cent of all births, deafness for 0.05 per cent, blindness for 0.02 per cent and albinism for 0.01 per cent. Others, such as Tay Sachs disease, are much rarer but occur more frequently in some populations than in others.

One of the more accessible demonstrations of genetic adaptation concerns blood groups. Polymorphism in blood groups is just one of the many genetic devices that may help insure populations against disease. Often polymorphisms are spread widely across racial or continental boundaries. However, the number of genetic traits that are geographically clustered suggests that particular traits not only have a place of origin but may also have become concentrated because they confer an advantage in that particular place. Many of these foci are only recently discovered and their functions and correlations are still unknown; many may turn out to be neutral.

O is both the commonest blood group and the oldest. It is apparently the only one shared with another primate – it occurs solely in chimpanzees and humans. Blood tests on chimps suggest a frequency of about 15 per cent.

Although the adaptive value of blood groups is difficult to prove definitively, most haematologists consider that the great diversity of blood group genes has evolved, like most other structures, as part of our adaptive response to the many challenges of life, especially disease. One of the indications that blood groups may be involved in protection against disease comes from a correlation between low frequencies of blood type O and the areas in which epidemics of plague were most severe in the past. It is known that O groupers are more susceptible to plague, whereas A and B are less so. Tests have suggested that A and B are also marginally but measurably less likely to attract mosquito bites, so there may be additional advantages. People with the A blood group category are also statistically less likely to suffer from gastric ulcers than members of the O group. As if to balance the scales, A group members are more prone to smallpox and typhoid. In this way, genetic polymorphism has probably helped nuclei of survivors to rebuild populations after the many successions of devastating epidemics that swept populations away in near, but seldom total, annihilation.

How do genes operate to combat disease? At least one adaptive advantage that could possibly be linked with a specific gene has been identified. The 'P1' gene (which is common among southern Africans, including the Khoisan) may help stimulate antibodies to certain worm infections, because helminth surface antigens resemble P1.

The geography of genetic peculiarities and of the diseases they may be related to is a central preoccupation of contemporary medical research. The tropical world offers numerous examples of dangerous diseases and parasites with restricted distributions. Swamp fevers, forest viruses and agues are endemic to parts of Brazil and Africa. Large territories that might otherwise have been attractive used to be empty of people, either because of traditional taboos or because previous populations had died out. By contrast, there are regions where a large human population can exist only because a genetic adaptation has been acquired after persistent exposure to lethal disease.

Of the many indications that people, like other animals, have been selected for resistance to diseases, none are more compelling than the five genetic conditions (and there could be more awaiting discovery) that give protection against malaria. Malaria has probably become a much greater menace to settled agricultural societies than it was to nomadic foragers. Nonetheless, the evolution of genetic protection against malaria is one of the outstanding confirmations that adaptation is a reality. When the total ranges of these five conditions are plotted on a map of the

Distribution of falciparum *Malaria and some genetic buffers.*

Old World and compared with the distribution of *Falciparum malariae* there is a very close fit.

In spite of massive antimalarial campaigns millions of people die of this disease every year. When people were fewer, more widespread and mobile it would have been more difficult for the mosquitoes to maintain the cycle of infection, so it is possible that malaria was originally restricted to a small and possibly nonhuman focus.

The cause of malaria is a protozoan called *Plasmodium*, which is transmitted by female mosquitoes (which need a blood meal to fuel their breeding). By sucking *Plasmodium* in and injecting it out, female mosqui-

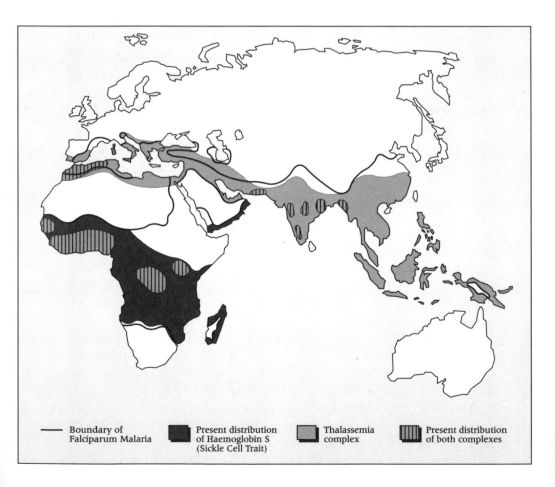

| — Boundary of Falciparum Malaria | ■ Present distribution of Haemoglobin S (Sickle Cell Trait) | ▨ Thalassemia complex | ▥ Present distribution of both complexes |

toes of the genus *Anopheles* quickly pass the parasite around and the *Plasmodium* then multiplies in the bloodstream in a succession of population explosions, each of which is marked by a high fever in the host.

When northern Europeans first started trading with west Africa it became known as the 'white man's grave'. Coming from outside the malaria belt they had no genetic protection against either malaria or yellow fever. The local people, on the other hand, had a battery of genetic buffers. Best known of these is sickle-cell anaemia, the genetically controlled deformation of blood cells. Sickling is one of several genetic peculiarities in the blood that are known to give some protection against diseases or parasites that infect blood. It is thought that the infective agent is less able to support itself on the abnormal blood. This could explain why carriers of the sickle-cell gene HbS are at an advantage over people with wholly normal blood (HbA) in malarial areas. The carrier has one normal and one abnormal gene (HbS/HbA), whereas the homozygous sickler has HbS/HbS and usually dies in childhood from acute anaemia. Normal blood types are at a reproductive disadvantage, because even if they survive repeated attacks of malaria, high temperatures and fever attacks in the men overheat their sperm and make them less fertile. In areas where anything up to 100 per cent of the population is infected with malaria, sickle-cell carriers (but not homozygotes) live longer, are more fertile and have more live children.

Plasmodium in the blood stimulates the production of antibodies in anyone, but most of these are useless and recent research has suggested why. It is thought that the parasite is protected by a protein coat that can be identified only by genes of a specific molecular structure that will bind to the *Plasmodium*, thereby mobilising the defences of the immune system.

One mechanism for an effective immune response depends on the highly variable human leucocyte antigen (HLA) genes on chromosome 6. Two genes of this complex have been found to protect west Africans from malaria, and it has been calculated that at least 2,500 generations subjected to malarial selection are required to explain their present prevalence. The protective HLA genes are common in many equatorial Africans (nearly one in two Nigerians have them) but are extremely rare among non-Africans. West Africans who have not inherited one combination (or haplotype) of HLA genes are more prone to severe malarial anaemia, while those suffering from cerebral malaria also tend to lack HLA genes.

West Africans have other mutations that offer some protection, Duffy African genotype Fy Fy increases resistance and, if this were not enough, many west Africans have the G6PD deficiency trait. This is an X-chromosome-linked inherited condition, which also occurs around the Mediterranean and is widespread throughout the tropical Old World, especially in the Far East. In Southeast Asia another genetic malarial buffer is haemoglobin E. Thus at least five mutations are known to have taken place in the red cells of tropical peoples. For the Danish sailors and English traders dying in eighteenth-century Accra and Calabar, a lack of adaptation was the ultimate reality that distinguished them from the Africans around their bedsides.

The capture and transportation of west Africans across the Atlantic as slaves is well enough known. It not only represented a monstrous conversion of people into a commercial commodity, it was also the exploitation of biological 'fitness': very large numbers of Africans for the profit and comfort of a small number of Europeans. Today the descendants of those slaves in the Caribbean undoubtedly retain their physiological advantages, such as superior resistance to skin cancers and sunstroke. What has probably been a more decisive advantage, especially along the 'fever coasts' of Caribbean Central America, is the barrage of malaria-resistant genes mentioned above. For the ordinary observer lacking both blood-typing kits and a genetic training the main signal of imported biological fitness in the Caribbean today is the prevalence of 'an African appearance'; yet appearances can be deceptive.

In the Caribbean and Central America the discovery that predominantly Amerindian genes can shelter behind an African appearance has intrigued several geneticists. A team from the University of Kansas first investigated this phenomenon in Tlaxcala, Mexico, and then focused on the 'Black Caribs' of Belize and St Vincent.

The island of St Vincent was first colonised by Arawak fisherfolk. Later they were subjugated by Caribs from South America. Whereas most Caribbean islands were annexed by European powers not long after Columbus's arrival, St Vincent escaped their interest until 1668 when a British expeditionary force grabbed it. However, this was not the earliest external influence. African slaves that escaped from neighbouring islands and others seized by the Caribs during raids were adopted or enslaved by the St Vincent Amerindians. From some time in the sixteenth century onward, the two peoples mixed to form what were first called 'Black Caribs' by an English visitor, W. Young, in 1795. Speaking the island Carib language and culturally Amerindian, this hybrid society was nonetheless already noted as being negroid in appearance in the eighteenth and nineteenth centuries:

> A Carib belle may even be seen in a Crinoline . . . few of the original colour now exist: the greater number being called Black Caribs. These last are of mixed race who in physical characteristics are rather African than American [Chester, 1869].

The Kansas team gene-typed their descendants in Sandy Bay, a Black Carib enclave at the mountainous north end of St Vincent. They estimated 43 per cent of Amerindian-type genes, 16 per cent European and only 41 per cent African-type. If a decisively African appearance can be the outcome of a mere 400 years and 41 per cent mixture it says a lot for the external visibility of this gene package and its implied contribution towards survivorship.

The Sandy Bay enclave has survived two major volcanic eruptions, several hurricanes, epidemics of smallpox (1849), yellow fever (1852) and cholera (1854). They are also the remnant survivors left behind after an almost total uprooting and dumping of the St Vincent Black Carib population on a small island off the coast of Honduras in 1795. Ironically the ship that carried them was called HMS *Experiment*.

Today the outcome of that experiment is some fifty-four communities strung out along the Central American littoral between La Fe and Stann Creek. In less than 200 years mainland Black Caribs have swelled their numbers to 80,000 without losing their cultural identity (they are known locally as Garifuna or Morenales). However, they have been inserted into much more extensive coastal populations with even more complex racial histories. There are the Meskitos, another Afro-American group, there are displaced Haitians and Creoles (a mainly Afro-European mix) and various Maya/Ladino or Euro-Amerindian 'hybrids'. Some measure of the genetic variety achieved by all this mixing emerged from the Kansas study: for example, forty-two distinct types of Gm blood group, compared with an average of eleven in Europe. What most of these coastal people have in common is a greater or lesser proportion of African-type genes but a decisively African appearance. Their dominance all along the Caribbean seaboard and the Gulf of Mosquitoes is an expression of hybrid vigour and adaptation by natural selection.

A lack of any resistance to widespread human diseases is most evident among previously isolated island communities. In the Central Pacific entire populations died of measles and other diseases that took quite a mild course in the continental sailors and adventurers that introduced them. Many islanders appreciated that foreigners were 'the plague' and attempted to repel them. It is quite likely that similar forebodings arm the fierce little communities that still exist on North Sentinel and South and Middle Islands in the Andamans. Up to the present they have avoided and repelled contact with the outside world while their relatives on the other islands have died out, largely, it is thought, through infertility brought on by syphilis.

This venereal disease is generally thought to have been endemic in the New World. Folklore has it that the sailors on Christopher Columbus' first voyage brought it back to Spain. Its rapid spread through Europe, as a very severe and apparently new disease, is well documented. At the time, traditional European enmities were invoked to blame old rivals for its introduction, and the 'Spanish pox' became English, French or whatever, according to the invective of the nations concerned. Significantly, epidemics seem to have had negligible effects on Amerindian people, implying that they had a longer history of familiarity with syphilis. By now most peoples have had more than a century of exposure and responses have tended to even out and have become less acute.

Lack of exposure, leading to a greater susceptibility to diseases, need not involve total isolation. At the turn of the century, when persecuted Jewish refugees escaped from their overcrowded ghettos in eastern Europe to settle in equally crowded new cities in the USA, many were tubercular. Many more had acquired a resistance to tuberculosis that was born of several centuries of ghetto life. In New York, Chicago and many other cities they were joined by large numbers of rural Irish immigrants. The Irish proved to be much more susceptible to tuberculosis, probably because their own contact with the disease had been less intense and sustained. Although health care helped curb the epidemic, a naturally acquired resistance may have helped hasten its decline.

Only a minority of diseases require genetic alteration to combat them and the same is true of environmental stresses. A great part of acclimatisation is due to simple exposure and training, preferably from childhood and, for many situations, this overrides built-in genetic endowments. For example, local populations have adapted to very high altitudes in the Himalayas, Andes and Alps. In each instance these otherwise very different peoples have adjusted their blood chemistry, circulation and breathing patterns in almost identical ways.

Numerous tests on human physiology have shown that a fit person performs much better in almost any measurable function than one out of condition. The scope for any normal person to train or improve their fitness is therefore very great. This capacity for meeting the many physical challenges of life is adaptability. Adaptability is shared by the whole of humanity and, today, it seems to dwarf the very real physiological differences that exist between different groups of humans. But adaptability is not the same as adaptation, with all its clever genetic adjustments.

Nonetheless, where coping with high altitudes and a shortage of oxygen is concerned, the boundary between adaptability and adaptedness can be less easy to draw. High altitudes pose many hazards, the principal being the thin air. As a result mountain people tend to develop large lungs, because they must take in more air to get the same amount of oxygen. For a pregnant woman the shortage of oxygen can be critical, since she has to provide for two bodies. Any impairment of her lungs through influenza or bronchitis is likely to affect or abort her unborn child. As a result, fertility, as expressed in birth rates, is rather lower at higher altitudes (notably in Peru where it has been measured) than in the foothills and at sea level.

High-altitude populations are short in stature, big-chested, suffer more from bronchial and pulmonary illness but have substantially healthier hearts and circulation systems than their lowland relatives. When they come down off the mountain and have their capacity for exercise compared with that of lowlanders, they often perform better. The amount of air breathed in by an alpine-adapted person can be as much as 40 per cent greater than for a similar person at sea level. Less obvious is the fact that the number of the smallest blood capillaries in the lungs, brain and muscles increase in alpine people. The net result is a great increase in the efficiency with which oxygen reaches tissues.

Understanding that people have significant peculiarities, both as individuals and as populations, has immense practical relevance for medicine. It also has many social and political implications. But there is also the deeper challenge to our imagination. By far the greater part of what we are, as both adapted and adaptable beings, has its origins in our ancestral foraging existence.

7: Eve's Descendants

Just as our view of the world and our origins are shaped by our circumstances so finding a way of explaining foreigners and defining their differences is the beginning of taxonomy. All taxonomies reflect the limitations of their inventors and they incorporate the mode in which the inventors structure their knowledge. Thus Carl von Linné's systematics sprang from biblical assumptions, in which an unknown (but divinely predetermined) number of fixed types had existed since the Creation. This was how he put it:

> I distinguish the species of the Almighty Creator which are true from the abnormal varieties of the Gardner; the former I reckon of the highest importance because of their author, the latter I reject because of their authors. The former persist and have persisted from the beginning of the world.

In classifying animals, Linnaeus (his Latin name) used obvious properties of shape and colour. There was no departure from these simplicities in the way he colour-coded humans into four varieties – white Europeans, yellow Asiatics, black Negroes and red Amerindians. The absurdities of this system now only survive in outmoded forms of speech but the attachment to types and typology is still very much alive.

Eighteenth- and nineteenth-century Europeans were not only burdened with typology, they also sought the 'ideal' (later reaccredited as the 'average'). Classical concepts of beauty infused this search and led at least one anthropologist, Kretschmer, as recently as the 1930s to seek out 'the most beautiful' specimens he could find of human body form to typify his so-called 'constitutional types'.

It was within similar idealistic traditions that, in 1770, the physician Dr Johann Blumenbach invented the term 'Caucasian' to describe what he considered to be the 'type' of western Eurasian man. This quixotic choice of a name derived from his possession of a skull that had been unearthed near Ararat in the Caucasian Mountains. Blumenbach thought this skull approximated to one of the five perfect racial types he thought he could recognise. In spite of genetics having shown that 'Caucasian' has no real meaning it is still used to describe a wide variety of western Eurasian people and their descendants.

This catch-all term describes a former population block the supposed boundaries of which are particularly fuzzy. It is essentially a waste-paper-basket category to hold all the people west of the 'Mongoloids' and north of the so-called 'Negroids'. In spite of fuzzy outlines, the recent distribution of 'races' was so neatly coincidental with countries or continents that the possibility that these 'types' might have been mixing their genes for thousands of years has remained difficult for people to accept.

The assertion that all members of the human race must in some sense share their ancestry with modern Africans was first greeted with incredulity. It was an idea that did not go unchallenged and in a not-too-subtle hint that the idea was mythology its opponents dubbed it the 'Garden of Eden' hypothesis.

The name was actually quite apt, because an idea first formulated by palaeontologists on the basis of fossils soon found some corroboration from a wide-ranging analysis of human mitochondrial DNA (mt DNA) by molecular geneticists from Berkeley, California. Rebecca Cann and her colleagues postulated that all Modern humans derived from a single African 'Eve'. Furthermore, calibration of their molecular clock suggested that there had been a combination of genetic innovations sometime between 140,000 and 290,000 years ago, whereby 'Eve' in biblical fashion begat the human race. This very approximate estimate of time has equally rough and ready corroboration from the fossil record.

Cann's original article in the science journal Nature drew a variety of critical responses from her fellow geneticists. Both the methods used to map and measure differences and calibrating molecular clocks are matters of controversy. Naruya Saitou and Keiichi Omoto firmly reminded the readers of Nature that data from mt DNA are not enough *on their own* to establish when or where human populations diverged. Professor Alan Templeton went further and thinks mixing in early populations may have so scrambled DNA that a true reconstruction may *never* be possible. Fortunately there are other sources of information to corroborate the basic correctness of the 'out of Africa' or 'Garden of Eden' hypothesis. Scepticism about this hypothesis hinged partly on the very real difficulty of explaining how 'generalised' people such as the Mediterraneans and the Indians could be derived from such 'specialised' people as modern black Africans. It is not generally realised that modern Africa and modern India are two parts of the world where humans are naturally at their most diverse; they are far from homogenous, and Africans are genetically more varied that any other people. The main reason why non-Africans fail to appreciate this is their naive reliance on a single characteristic or signal – colour.

The churning of human genes that has taken place all through the tropical Old World defies easy untangling, and one mistake has been to suppose that deeply hybridised people can provide easy genetical or anatomical bench marks that will allow twentieth-century relationships to be sorted out. Some critics of 'out of Africa' suffered from this handicap. They could not match their conception of modern Africans with the Eves and Adams who went forth, multiplied and inherited the earth.

Genetics is a science that is still in its infancy but it has already provided many insights into human evolution. For example, there is a strong implication in the work of several molecular biologists that 'races' have diverged relatively recently. In 1980, B.D. Latter compared protein polymorphic variation within and between populations of major human subgroups. He found that differences among *individuals* of the same racial group accounted for 84 per cent of the variance. In other words, racial differences are only one sixth as significant as individual differences.

When Rebecca Cann and her colleagues put forward the idea of an African Eve the cornerstone of their argument was the extraordinary diversity of mt DNA that had been found in San women from a small area in southwestern Africa. Their conclusion was founded on the fact that mt DNA passes down only the female line of inheritance. This is because male mitochondrial genes are always lost when the sperm penetrates the ovum. As the 'energy batteries' of a cell the mitchondria carry their own complement of thirty-seven genes quite seperate from the cell nucleus.

While harmful mutations simply eliminate their bearers there are sequences in mt DNA that accumulate neutral mutations at a rate that is thought to be constant and calculable. For genetic material that evades recombination, such as mt DNA, there are very important clues to past relationships and connections and, of course, a mechanism for setting a molecular clock (so long as the correct calibration could be found).

The timing of Eve's birthday was based on the prediction that mt DNA mutates at a constant rate of 2–4 per cent per million years in all vertebrates. Although these mutations accumulate as quantifiable measures of change, they are considered to be neutral and without influence on genetic coding.

Although the concept of genetic clocks is still controversial, there is a fair correlation between the percentage of mt DNA divergence within a discrete population and how long people are likely to have been established in a given area. For instance, a sample of thirty colonists in Venezuela were effectively identical in their mt DNA. A larger sample in Japan showed 0.26 divergence. Nine hundred and fifty-one Italians were 0.32 divergent, while the thirty-four San women from the Kalahari were found to have the highest level of sequence divergence yet recorded, 0.59 per cent. Other African samples also show higher percentages than European and Middle Eastern samples. Theoretically the greatest divergence would occur in areas where female lineages have stayed put longest. In such localities women would retain their highly differentiated DNA even if there had been extensive mating with invading males from other areas.

Such divergent mt DNA is not found outside Africa, so there is the strong implication that Modern humans remained in Africa for a long time before venturing into Eurasia. Mt DNA has been used to grow a genealogical tree of human divergence and differentiation, because no recombinations or segregation of parts is involved, just a steady and fairly rapid rate of change as it passes from mother to daughter. Every time a small group hives off to found a separate dynasty, the total variability within that population drops. Hence there are three very significant implications in the measure of 0.59 per cent divergence in only thirty-four women. One is that the San are the most stay-at-home people on earth. Another, that their female ancestors have always lived in fairly large numbers in southern Africa. The third implication is that San genes have been around there much longer than any other genes anywhere else on earth.

This, of course, does not preclude change and some genetic mixing with neighbours. That there has been long-term admixture between the

San and other Africans has been inferred from several blood groups, especially haptoglobin (Hp^1), which is at very high frequencies throughout tropical Africa. This blood group has 18–30 per cent gene frequencies among the San. On the other hand, three blood types are virtually absent among the San yet widespread in tropical Africa.

The modern Khoisan are known to be smaller than some of their antecedents (perhaps as part of specialisation towards a more restricted range and environment). In any event, a large number of fossils attest to a larger-sized but still recognisably Khoisan people having occupied the whole of southern Africa for many tens of thousands of years.

The oldest skull originally thought to have Khoisan features is a broken fragment of face and cranium which was discovered in South Africa at Florisbad. Although undated and incomplete, this skull is now thought to have closer affinities with fossil skulls from Tanzania, Sudan and Morocco than with modern Khoisan. What is really exciting about the Florisbad and other ancient African skulls is not just their remote links with a modern African people but that those respects in which they are different from the Khoisan take them closer to modern Australians and early Europeans. In other words, the Khoisan, like everyone else, have continued to evolve but they have a direct line of descent from those first Africans who gave rise to such different and distant peoples as Australian Aborigines and Europeans.

Genes that are not involved in coding, such as mt DNA, are more valuable as indicators of relationship than those that are subject to natural selection. The frequencies of the latter can change very rapidly within a population and so cannot tell us anything about the distant past. However, the Y chromosome is from a non-coding region and studies on variation in this chromosome are also unequivocal in suggesting African origins.

Khoisan genes are still widely distributed in southern Africa. Their particular gene peculiarities are present, to a greater or lesser extent, in most southern African people (except the most recent immigrants, such as the Europeans, Asians or Ovambos). The principal genetic marker of the Khoisan in Gml, 13, 17. This allele's variable frequency in all Africans in southern Africa suggests that a large proportion still retain some Khoisan genes. Among the Xhosa the ratio is as high as 60 per cent, among the Zulu 45 per cent and as far away as Malawi 11 per cent have

The reduced size of modern Khoisan, skulls
and reconstructions. (left) A modern khoisan.
(centre) Late Stone Age fossil, Khoisan skull
from Fishoek, South Africa. (right) Late Stone
Age fossil skull from Homa Bay, Kenya,
resembled a heavily built Khoisan.

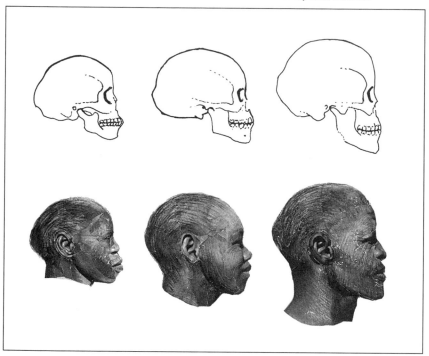

this Khoisan gene. Of their closest neighbours only the Ndebele, Ovambo
and southern Angolan tribes seem relatively unmixed, which suggests
that these people have come down from equatorial Africa quite recently.

Today, the San people of Botswana, Namibia and Angola number only
about 70,000 and their way of life is constantly being eroded by outside
pressures, such as being recruited into the army, mines or farms. When
a 'new Bushman tribe' was said to have been discovered in the central
Kalahari Desert, it was found that all its fittest young men had worked
for some years in a particular gold mine outside Johannesburg – such is
the lure of the cities!

The two most damaging intrusions into San life in recent years have
been the undeclared war between South African forces and nationalist
guerrillas in Namibia and World Bank-financed cattle schemes in Botswana. Both these developments have reduced these independent
communities to servitude, and have accelerated the process that overtook the San and Khoi in South Africa.

Modern distribution of the San with estimated percentage frequency of the 'Khoisan gene' Gml, 13, 17, in neighbouring groups (from Jenkins, 1970).

The Khoi were the first South African people that encountered European merchant navies. They were herders of sheep and their territories centred on water sources. Initial contacts stimulated entrepreneurial activities among the Khoi, who helped supply meat and water to ships at anchor in Table Bay. At least one Khoi from the Cape of Good Hope went to Europe in the seventeenth century to learn European business practices, but when Dutch settlers took over the Cape in the eighteenth century the Khoi were soon reduced to becoming 'Hottentot' slaves and servants. Later, with the growth of cities and towns, their identity became still further degraded. Now a minority, they joined other minorities in the urban labour markets where historical roots and human dignity were subsumed in the meaningless name of 'coloured'. Although the way in which their independence and identity were lost has been peculiarly degrading, their incorporation into other social systems is neither unique to them nor was it something new in their own history.

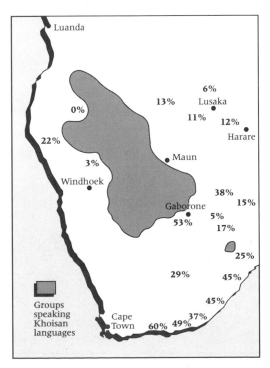

The erosion of the Khoisan.

The studies of Khoisan gene frequencies have demonstrated that the original inhabitants of this very substantial region of the earth contributed their genes to virtually every invading group and that this process of mixing continues unabated. An inevitable corollary of this is that the Khoisan themselves were far from 'pure' and cannot be taken as the unadulterated aborigines of southern Africa. Nonetheless, people of Khoisan ancestry are certainly among the closest we shall get to the direct line of descent from our common grandmother.

We cannot expect Eve's features to be reflected directly in Khoisan faces but there are some very persistent and widespread traits which turn up in surprising places. Their short, broad and flat upper faces with oval outlines and narrow, heavily lidded eyes are often likened to the Chinese but closer inspection and detailed measurements make it clear that this is a wholly superficial resemblance. Nonetheless, these two modern peoples may share a small subset of facial features characteristic of one echelon of Eve's offspring. A perceptive student of the Khoisan, William Howell, has suggested that these and other typical features might be due

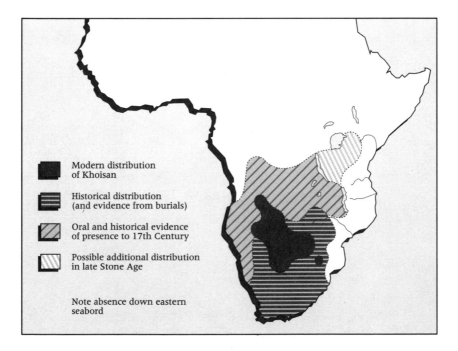

Modern distribution of Khoisan

Historical distribution (and evidence from burials)

Oral and historical evidence of presence to 17th Century

Possible additional distribution in late Stone Age

Note absence down eastern seabord

to a reduction that 'may occur both in general size of the skeleton and skull and particular features, like the brows or the face, by small-scale evolution or natural selection of some sort'. While it is quite likely that some of the respects in which modern Khoisan depart from their larger predecessors could be influenced by later admixtures, others may, indeed, represent 'small-scale evolution'.

Several studies of change in later prehistoric populations have found much variation in body build but a general trend towards more lightly built people. Although some aspects of this are clearly reversible (dramatic increases in average body size and bone development are possible over short periods), regional characteristics remain recognisable in spite of these changes. The problem with early Africans is that there are still too few specimens to establish the limits of variation but the early fossils do tend to bear out a presumption that the earliest populations diverged rather slowly, only gathering momentum and becoming more distinct as they multiplied in numbers and adapted to ever more variable habitats over an ever larger range. Such conservatism would not have precluded considerable individual variation in size and physical development.

In the case of the Khoisan's shrunken range, their presumably adaptive changes in appearance could have been specific to the local environmental demands of southwestern Africa. However, if they do retain older, more generalised tendencies, then their rounded flat features may belong not just to the Khoisan but to a much broader guild of Eve's descendants, who might respond in similar ways to 'small-scale evolutionary change', especially when body sizes are reduced. In this context, Khoisan-like skulls (notably Qafzeh IX), dated at about 110,000 BP and found at Jebel Qafzeh in the Levant, may be less surprising than at first glance. More difficult to explain is another apparently 'Khoisan' skull from Liujang in China, which may be as recent as 20,000 BP.

Modern Khoisan are notably broad-skulled (brachycephalic). This trait too, like small size, may be linked with their later, more concentrated distribution. At Gwisho, in Zambia, thirty-three late Stone Age skeletons, all of undoubted Khoisan type, were larger-bodied than modern San, and many had longer (less brachycephalic) heads. The Gwisho men were also much taller than the women, a dimorphism that is not typical of modern Khoisan. The longer heads and the greater differences between the sexes that are visible in the Gwisho Khoisan type

263

Khoisan. Collage of four individuals. Infant, young adult and elderly females. Middle-aged male (Photo R.White).

resemble some traits typical of northeast Africans. Yet another skull, from Homa bay in Kenya, combines similar traits (see p. 260).

The great Lakes region of East Africa and Ethiopia probably represents the hub from which modern humans radiated out. Today there are many distinct peoples in Kenya and Ethiopia, most with long skulls, relatively prominent noses and chins and of tall athletic build. The Khoisan could, therefore, retain traces typical of the 'southern branch' of Eve's descendants while, north of the equator, there could be some sort of continuity between humans of the Omo I type and a very conservative, long-lasting lineage that was called Cro-Magnon in Europe and Mechta Afalou in North Africa.

Compared to the Mechta Afalou or the Cro-Magnons, modern Africans from the northeast are slight and slender, but there is no reason to suppose that they do not trace at least some of their genes back to the earliest inhabitants of the northern half of Africa. Today, these peoples tend to be placed into large linguistic categories: Cushites in the Horn of Africa, Berbers (Touaregs) in the northwest, Chadics, Nilotics and Nilo-Saharans in between.

In a study of population change in northern Africa over the last few thousand years, M. C. Chamla found that people of the hefty Cro-Magnon type were widespread up to the end of the Neolithic. About 10,000 BP 'proto-Mediterraneans', people who were clearly ancestral to today's North Africans, begin to occur in the same sites as the Cro-Magnons. However, these people were just as robust as their predecessors to begin with. So, here again, a variation of Howell's small-scale evolution might have been at work. Chamla also found that as the Cro-Magnon declined, the Mediterraneans became more lightly built but without much alteration of basic type.

Today the Middle East is dominated by Mediterraneans that belong to Semitic language groups. The Mediterraneans are commonly described as part of the highly diverse 'Caucasian' category. In spite of their variety and a degree of diversity that is greatly accentuated by cultural and national differences, the contemporary range of 'Mediterraneans' or 'Caucasians' extends from the Atlantic to the Bay of Bengal.

In India western invasions have come in successive waves and the layers of past movements could be expected to have all but obliterated the earlier populations of India. As the first really extensive area of prime habitat that the early Africans encountered as they expanded eastwards, India should also give signs of its past, if not in its living populations at least in its archaeological deposits. The churning of peoples makes any study of their genes extremely complex. Bones and organic artefacts are also extraordinarily scarce in India; many millennia of dense and widespread occupation and movement may have combined with an erosive climate to obliterate much of the past.

Nonetheless, there are both archaeological remains and living people in southern India that have enough similarity to early Africans to corroborate that India was indeed host to the earliest expansions out of Africa. Late Iron Age and Neolithic skulls from Bayana and Sailkot in Kerala and Adithanallur in Tamil Nadu have well-developed brow ridges, receding foreheads and the over-all heavy build that is typical of all early Moderns and was labelled 'Proto-Australoid' by the excavators.

Few south Indians could be described as heavily built today. Like humans almost everywhere they are much lighter in build than their predecessors. Nonetheless, here, at the southern tip of a busy and vastly populous subcontinent some conservative lineages might have maintained a degree of genetic continuity. The situation could be comparable

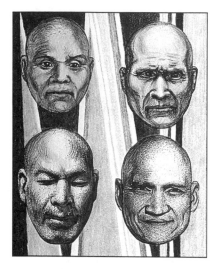

East Asian Islanders. (top) Hokkaido (Ainu) middle-aged and elderly males. (below) Hokkaido (Ainu) middle-aged male. (below right) Taiwan middle-aged male.

with that of other continental cul-de-sacs, in that older populations are less influenced by later ebbs and flows at the extremities of a landmass. Against this must be balanced all the coastal comings and goings along the Indian Ocean shores. Some of the southern Indian tribes are foragers or part-foragers, notably the Gond, Chenchu, Juang and Yerukala, and these tribes tend to keep mixing to a minimum. A large measure of distinctness among nontribal communities has also been maintained in India through the extremely durable social structures imposed by caste, religion and occupation.

Because very early populations have had the most time to spread and adapt to difficult and marginal environments it is not surprising to find relict populations at the peripheries. The most peripheral and isolated, but still habitable, regions of the Far East are the islands of Hokkaido and Sakhalin. There the northern Japanese or Ainu survive in very small numbers, rapidly losing their identity as they merge with the Japanese mainstream. Japanese genetic make-up differs from the Chinese in ways that show that most of the former retain some Ainu inheritance. The Ainu are directly descended from the Jōmon, a name used to describe a culture that had developed throughout the Japanese islands by about 12,000 BP. The Jōmon were probably isolates of the earliest Palaeolithic humans that had established themselves in the Far East. The isolation of Jōmon and Ainu from external gene flow may help explain their many conservative features. Of all Far Eastern people they retain the most of the generalised characteristics that would have come into the region via India. They tend to have heavy brow ridges with a depression at the root of the nose. Their heads are long with a short face, noses tend to be broad and the eyes of males are far less prone to the narrowing that is typical

Australians. (top from left) Queensland middle-aged female; Tasmanian middle-aged female; Queensland youth; Arnhemland female; West Australian elderly female. (bottom) Queensland adult male; South Australian elderly male; Central Australian elderly male; Arnhemland adult and aged males.

of all other Far Eastern people. Their abundant body and facial hair is wavy and dark brown rather than black and straight. They are powerfully built but generally very short. Genetically they largely resemble southern Japanese.

Dr T. Kanaseki and Dr K. Hanihara have shown that modern Japanese people are probably the product of very late invasions of mainland people (mainly Yayoi from Korea), who entered Kyushu about 2,300 years ago and intermixed very extensively with the native Ainu. Even today, Japanese from the west are more 'Yayoi' (higher foreheads, narrower noses and flatter, rounder faces) while eastern Japanese are sometimes more 'Jōmon' (shorter in stature, with long heads, lower foreheads, broader noses and browner hair).

The proposition that very early populations are the most likely to find their way to, and then adapt to, the furthest habitable lands that they can reach applies not only to the Ainu but also to the Australians. The Ainu represent relics of an early northeastern expansion at its outer limits, and the Aborigines are relics of a similar southeastern expansion. There is, however a big difference. The southern pathway to Australia had to cross the equator and a multitude of islands. The climate, the way of life and the technology necessary to accomplish this amounted to a sort of cultural and physiological filter through which the proto-Australians had to pass.

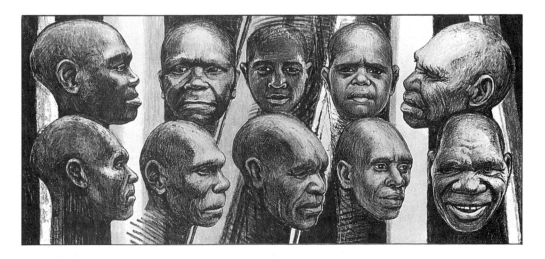

The contrasts between prehistoric and Modern people – heavy and bony on the one hand, and light and more juvenile-looking on the other – are manifested in Australia in their extreme form, both in fossil and living people. In both cases extremely powerful and massively built people exist on the same continent (and among Moderns) at the same time as small, delicately featured people. The former are typified by the southern Australians, who live in an environment with very hot dry summers and cold winters; the latter by northeasterners living in dense forest under a very wet climate.

In an early study of Australian Aborigines, the physical anthropologist Birdsell stated that he could find metric and other physical differences between tropical populations in the northwest and those dwelling in the temperate southeast. The former he related to the Vedda of Ceylon, noting tendencies to tallness, flat noses, very large brow ridges and narrow craniums behind broad faces and the fact that they are very seldom bald. The southerners are less black, have big chests, are hairy, have higher bridges to their noses and older men tend to go bald. These traits he called 'Murrayan' and he declared that, in all measurements except stature, they were closest to the Ainu. Birdsell also erected a third type to accommodate the north Queenslanders, where close contacts with New Guinea are known to have a long history, confirmed by genetics and cultural similarities.

These differences were the premise for an elaborate theory of three waves of invasion from Asia. There is a simpler explanation. The Vedda, Ainu and Australian Aborigines all have a common origin, although each has continued to evolve in relative or complete isolation. The northern Aborigines and the Vedda have a tropical environment in common, while the Ainu and the 'Murrayans' live at much the same degrees of latitude north and south of the equator. In spite of all Australian Aborigines having passed through the Banda equatorial seashore filter, the more severe disciplines of a tropical life have been lifted for long enough to permit some generalised traits to be retained or to reappear in the south. These traits may or may not be adaptive but are common to both 'Murrayans' and Ainu, perhaps because both have been relatively isolated from gene flow and both derive from a similar conservative stock.

The general consensus at present is that all Australian Aborigines owe the greater part of their make-up to a rather small number of ancestors. But how many? Eve in Africa and a later Aboriginal Eve in Australia are two rather different propositions. Nonetheless, with his tongue in cheek, John Calaby, an Australian zoologist, floated her alone and already pregnant on to Australia's virgin shores. Calaby's humorous contribution to an early discussion on the origins of Australians was the most economic of all explanations and it gave specific focus to the question of how many people it takes to populate a continent.

In 1986, Mark Stoneking addressed this question with an analysis of mt DNA and suggested that there were fifteen to twenty lineages derived from at least that number of colonising females. The descendants of these founders would have spread over the whole of Australia where they were buffered against further major incursions by several barriers. There was the sea and the forbidding nature of the terrain at the normal point of entry in the northwest, but most important was their own numbers. Stray arrivals would have been unlikely to have much ecological space to expand into because a difficult terrain was already totally occupied, so that newcomers were either eliminated or genetically swamped by the much larger established population.

There have been various estimates of how many Aborigines occupied Australia at the time it was first settled by Europeans. One figure that is widely bandied is 300,000 at around 1790. More realistic estimates argue a figure in the region of 1.5 million. It is more than likely that in over 60,000 years of occupation human populations rose and fell many times, perhaps greatly surpassing both these estimates at one time or another. For hunters the initial resources of the continent would have been much greater than in 1790. The most effective way of living off them might have taken time to develop but intensive exploitation of large and vulnerable animals combined with climatic and fire-induced changes would have led not only to animal extinctions but also to very substantial fluctuations in human populations, as adjustments were made to new and very often degraded ecological regimes.

The genetic variety found in Aborigines has contributed to the development of elaborate theories and quite extreme positions; some postulating major incursions from various parts of southern or eastern Asia and others claiming total isolation. The variety of Aboriginal genes could have four sources: new mutations, random genetic drift and mate exchanges in isolated communities, localised natural selection and new genes from outside. Genetic profiles have tended to confirm that the Aborigines passed through a demographic bottleneck by showing a smaller spectrum of types than their neighbours, but they also confirm that they have had ample time to evolve their own regional peculiarities.

Foragers live in small and dispersed units and this would have had its own influence on genes. For example, most Aborigines lack the B gene in the ABO blood-group system. Another, the human leucocyte antigen system, links some Australians with New Guineans, but at a rather distant level. Another clear sign of local affinities is the unique mutant peptide B6 shared by a small group on the Gulf of Carpentaria and by an equally small group among the Asmat in southwest New Guinea. Affinities between coastal peoples either side of a shallow 'temporary' sea are to be expected but it is striking that south of this coastal fringe there are decisive differences between Aborigines and Melanesians.

The differences between New Guinea and Australia would have been greatly accentuated by the former's more frequent and influential contacts with Southeast Asia. This contact undoubtedly helped to get agriculture going, introducing both the pig and taro as major staples at about 10,000 BP. Agriculture can support a much higher density of people and the total for New Guinea, even in prehistoric times, may have been in excess of 2 million. The much larger landmass of Australia is unlikely to have reached this figure and the combination of small numbers and isolation would have had a profound influence on gene patterns.

The main difference between Australian and Melanesian environments is that the former is predominantly arid and open with foraging and fishing as the only modes of subsistence, whereas the latter is mostly densely forested with the majority of contemporary and recent people involved in agriculture, fishing or both. In spite of self-perceptions in which local tribes think of themselves as unique in tongue and origin, the overall pattern is of many layers of inter and intra island movements from Banda times onwards and much more recent but selective incursions by Austronesian speakers that were often more Polynesian or Southeast Asian than Melanesian. There are some very unusual-looking people on particular islands or in localised valleys of the highlands; large hooked noses occur in regions of the latter, blond hair or narrow faces

The diversity of Melansians. A cluster of villagers from the remote Gogol Valley display the variety of physiques typical of contemporary Papua New Guinea.

in the former. Few in the South Pacific would argue with the proposition that the islands' isolation has been the principal cause of local human diversity. Contemporary marrying-out is a mixer which is unlikely to be anything new but is another cause of diversity.

The immense variety of peoples, languages, cultures and gene types that occur in Melanesia has several sources. Firstly, there are the repeated immigrations, emigrations and interbreeding between highly mobile, sea-oriented peoples. Secondly, this process has been going on for at least 50,000 years, especially within the huge and dissected island of New Guinea. Time and the ever more complex permutations that go with it have multiplied the number of naturally diverse people by adding new hybrid groups. Thirdly, as their island realm became more and more far-flung, the ancestral Banda had already begun to differentiate even before Australia and New Guinea were reached, and distinct genetic strains would have appeared and gone on proliferating.

Some measure of the diversity of people and cultures in the South Pacific can be appreciated by the number of languages: there are about eighteen Papuan language groups ranging from Timor to the Solomon Islands and hundreds of Austronesian languages, which are late intrusions, dating only from the last 5,000 to 7,000 years. There are also many Australian languages which have been isolated for too long to bear any

Melanesians. (top from left) New Guinea (Gogdala) adult male; New Guinea (Huli) adult male; New Guinea (Highlander) middle-aged male; New Guinea (Telefol) adult female. (below) Solomon islander adult female; Vanuatuan middle-aged male, New Guinea (Kewa) adult male; New Guinea (Duna) adult female.

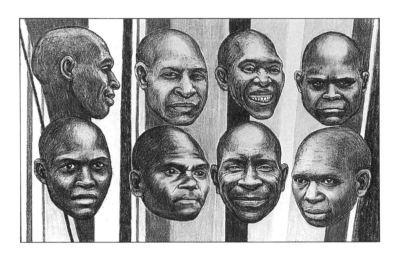

detectable relationship to the Papuan ones. The great linguistic expert Dr Greenberg holds out hope that some relationship may ultimately be traceable between Papuan, Andaman and Tasmanian languages. If Greenberg is correct this would imply that extremely isolated cultures might conserve linguistic patterns for many tens of thousands of years.

Among the many and far-flung populations of dark-skinned, spiral-haired people are the numerous relict groups commonly called Negritos (which is what the Spaniards named the black Filipinos). They occur throughout Southeast Asia and most of them now show some genetic resemblances to their Mongol neighbours, particularly those on or nearest to the Asian mainland.

In a wide-ranging study of six Philippine groups, a Japanese genetic survey suggested that the taller, darker Mamanwa of Mindanao grouped with similar minorities on Timor and the Moluccas. Further north, Negritos had fewer differences from the more Mongol majority around them, but there was a Mamanwa connection among the Agta and Dumagat easterners, while the Aeta and Batak westerners were different and a shared ancestry with the Semang in Malaya was suggested. These studies therefore confirmed that there has been a substantial divergence within the Negritos as well as much later mixing. Nonetheless, the researchers concluded that all have shared some ancient lines of descent from a generalised late Pleistocene Sundaland population. This conclu-

An Agta (Filipino) young man, Tumoy, with his bow and arrow and barkcloth loincloth (Photo J. Kamminga).

*Agta Nigrito, Tero, with two Tagalog visitors
from Manila. Baclai, N.E. Luzon, Philippines
(Photo J. Kamminga).*

sion is important because there have long been suggestions that the
various Negrito populations of Southeast Asia could be explained in
terms of convergent evolution. The argument was that certain unspeci-
fied and unknown factors might trigger genetic changes in quite widely
disparate localities. No such genetic mechanisms have been identified
and their existence is, to say the least, dubious.

Until recently the Agta of northern Luzon were nomadic hunters (but
with small territories). They fished in rivers and along the shore of Cape
Engano, dug wild yams and ferns, trapped game and hunted wild pigs.
It is some measure of their deliberate avoidance of outside contacts that
they continued to tip their arrows with bamboo rather than metal until
the middle of the twentieth century. Today they are victims of land-
grabbing and persecution.

Southeast Asians (Nigrito). (top from left) Philippine (Aeta) middle-aged male; Malayan (Batek) young adult female; Malayan (Semang) adult male. (below) Malayan (Semang) adult male; Philippine (Aeta) middle-aged female.

The black minorities in Southeast Asia are now being rapidly assimilated into their respective mainstream nationalities; Indonesian, Malayan, Filipino, Indian and so on. This is the sudden, final acceleration of a process that must have been continuing for thousands of years. The overall pattern is consistent with all these peoples having a common but very ancient affinity, which is their Banda inheritance. Not only did their widely scattered island existences encourage considerable divergence within the early Bandas but all have had later admixtures, some emanating from the mainland, some from other islands and all, to a greater or lesser extent, Mongol. These have been given all manner of names, most of them designed to fit in with a particular model of human diffusion or classification: for example, proto-Mongoloid, proto-Malay, proto-Polynesian, Austronesian and Deutero Malay.

What of the most westerly of all these island blacks? In 1973, Howells described these little-known people in the following words:

> The Andaman islanders of the Bay of Bengal give us pause – they seem to be different, though otherwise they would appear to be a natural extension of the 'Negrito' realm. They have their own language(s) and enjoyed an isolation of unknown length before modern times; and their diet of wild pigs and dugongs, among other staples, should meet the calories and protein needs

of much larger people. But they are Pygmies indeed, averaging about four foot ten inches and being very dark with woolly hair . . . the Andamanese, with their infantile faces and bulging, browless foreheads, look like little Africans. And statistically that is where their skulls belong. Andamanese crania group themselves with three African negro populations *more closely even than the African Bushmen* . . . They are more distant from any Pacific peoples; among these they most nearly approach the New Britain Tolai . . . I cannot explain the Andamanese and nothing is known about their past.

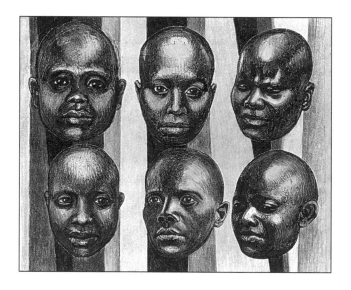

Howell recognised that the Andamese belong to an Indo-Pacific constellation of island blacks; he also saw that they are peculiarly different. They undoubtedly represent an isolated relict of an earlier, more widely distributed group. The observation that Andaman skulls group more closely with tropical Africans than with Khoisan Bushmen is consistent with their being the sole survivors of a distinct Banda group that probably centred on the coasts of Sumatra and both shores of the Andaman Sea.

The Vedda, as the earliest colonists of Sri Lanka and separated from India by a 60 km strait, might have preserved an isolation that was only slightly less prolonged and less complete than that of the Australian Aborigines. Vedda people have distinctive gene patterns (notably, very low Rhesus negatives) that confirm a very high level of sustained isolation. Like the Andamanese the Vedda may owe their survival as a distinct ethnic group to a prehistoric repulsion of all visitors and intruders. If so, coastal movements around the tip of India may well have bypassed Sri Lanka, thereby preserving this bastion of 'early Banda' against the 'later Banda' represented by the ancestors of the Andamanese.

Howells' comparison of the Andamanese with Africans and the Tolai has an important spatial dimension. Following the coasts of Burma, India and Arabia, the first landfall in Africa is 9,000 km away and the delta of

the Zambezi about 13,000 km. Following similar coast and island-hopping routes, Melanesians in small canoes had to cross much larger stretches of water to reach the Fiji Islands. They are nearly as distant from Andaman as the Zambezi, while the Andamans are about equidistant between the Horn of Africa and the Tolai's New Britain.

It is not only a question of skull shape, skin colour and hair type. The Japanese geneticists Naruya Saitou and Keichi Omoto have pointed out that at least seven different mt DNA genetic methods demonstrate a clustering together of Africans and New Guineans. Yet, they point out that analyses of nuclear DNA are equally unambiguous in clustering New Guineans with Australian Aborigines. This may be due to the different modes of inheritance between the two types of DNA. Another indication of relatedness between Melanesians, Australians and black Africans is the Transferrin mutant D, which is widely distributed in all three populations and is otherwise very rare.

As early as 1971, the geneticist R. Kirk pointed out that some people within these three distinct regional groupings could have shared common ancestors. In 1986, Wainscoat found two gene types in Melanesians which had previously been thought to be unique to Africa. Because Africa had been seen as the archetypal home of black people and they were 'discovered' and studied earlier, it has been almost automatic for scientists to see Africa as the source of such genes. However, when all the data are assembled, it is quite clear that the genes have flowed westwards and not eastwards. Considering that many tens of thousands of years later, after much further evolution and very considerable admixture with other people, these genes can still be traced in these small relictual populations as well as among the inhabitants of Africa, now numbering hundreds of millions, it seems likely that they confer decisive adaptive advantages.

A large number of Banda advantages may be concealed and operate in relation to disease resistance and high tolerance levels for heat and radiation stress, but there are two obvious 'western Banda' markers – a black skin and spiral hair. To this can be added a type of physiognomy that is generally described as 'negroid' – a short face, full lips, a fairly flat and broad nose – and although head hair is dense, bodies tend to be naked of all but the finest hairs. At first glance these characteristics are almost universal in Africa, but this is not the case; there are several broad regional patterns that suggest that Banda genes are superimposed upon at least two very much more ancient substrates.

One is the Khoisan. There are highly significant differences in the results obtained from mother-line mt DNA and those from nuclear DNA which mixes parental inheritance. The former lumps non-Khoisan Africans with several non-African groups, whereas the latter links the Khoisan more closely with Pygmies, Ethiopians and other Africans than with non-Africans. These differences have been taken to imply a very low mobility for female as against male foragers, which is probably correct. Another implication could be that there has been more genetic admixture, especially from males, in all non-Khoisan Africans. The main source of this input would have been from Banda people.

There are several conservative traits that seem to link the Khoisan with a much wider world than their immediate neighbours. Two features that turn up in widely different parts of the world are flatness in the face, especially across the cheeks and orbits, generally combined with a tendency for the incisor teeth to meet at more of an angle rather than vertically. The trait is known as 'shovelling' and occurs in parts of Africa and Europe and America, but most frequently in China. Both facial flatness and shovel teeth are typical hominid traits that have been retained in some areas more than in others. They may be further linked with a tendency for the bridge of the nose to be so reduced as to enhance the horizontal eyefold. This combination of features contributes to the stereotype of the Mongol or Chinese face and has led to the Khoisan being nicknamed 'Chinese Africans'.

The Khoisan are also readily distinguished from the main African groups by their click language. Bantu languages, by contrast, *sound* rather like some Papuan tongues. The Bantu and other related West African languages now dominate the whole of tropical and southern Africa. Their speakers are very heterogeneous and the spread of Bantu people has been linked with diffusion of the early Iron Age and forest agriculture. Bantu speakers have a much more extensive range now than two or three thousand years ago. Since languages readily cross ethnic boundaries it is misleading to equate a language group with ethnicity. Nonetheless, these languages have been said to have origi-nated in a broad band of country than runs northwest from Mozambique across the Zaïre basin to Cameroon. This is the primary zone in which the Banda–pre-Khoisan interaction took place. Among the most direct descendants of this first interaction are probably the various Pygmy people, such as the Baka, Aka and Mbuti. The taller agriculturalists of equatorial and western Africa have a complex history but can almost certainly trace their main line of descent back to the Stone Age foragers of central Africa called Lupemban or Tshitolian mentioned in Chapter 5.

The primary mix of all these people is Banda/pre-Khoisan. Further north, it is difficult to reconstruct the circumstances in which Banda influence might have spread. In any case a population that had already incorporated Banda genes would have encountered people who, like the pre-Khoisan, were aboriginal Africans. Although it is possible that these were still a single recognisable group, I think it more likely that they would have already differentiated into at least three types by the time Banda influence was established. Each could be surmised to belong to a distinct regional and ecological zone. The most identifiable of these would have been the Saharans. In the archaeological record these are called Aterians or pre-Aterians and the Berbers might be among their

The approximate distribution of modern language groups and their possible relationship to suggested Stone-Age differentiation of populations in Africa.

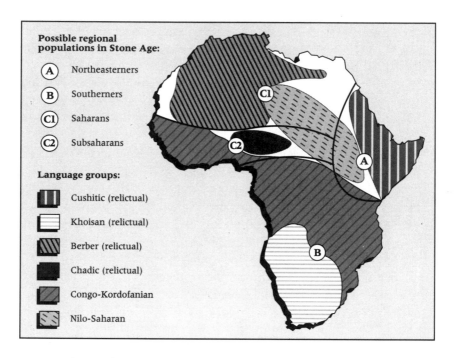

Possible regional populations in Stone Age:

(A) Northeasterners

(B) Southerners

(C1) Saharans

(C2) Subsaharans

Language groups:

Cushitic (relictual)

Khoisan (relictual)

Berber (relictual)

Chadic (relictual)

Congo-Kordofanian

Nilo-Saharan

present descendants. Banda admixture with this group was probably very delayed. Sandwiched between the Sahara and the equatorial forests was a broad band of prime habitat, much more like the southern savannas than the Sahara and possibly supporting a distinct population. If there are any genetic vestiges of this population today, they are likely to reside in the pastoralists found around Lake Chad and northern Nigeria, where a distinct and archaic language group still survives.

The uplands of Ethiopia and East Africa, surrounded by dry, open savannas, would have been the last region to receive Banda influence from Bantu or west African sources. That influence *has* occurred but so recently an alternative source of Banda genes is worth consideration. The Banda would have been exceptionally well adapted to exploit the hot, narrow shores of the Somali and Red Sea coasts even if at a very low density. But the genetic influence of such marginal and specialised groups would have been much slower to spread than in the south because their numbers would have remained so small. Nonetheless, if the Banda physiological advantages were as real as I have endeavoured

Northeast Africans. (top from left) Masai adult male and Juvenile El Molo male youths; Oromo middle-aged female; Samburu elderly female. (below) Masai elderly and middle aged males; Rendille elderly male; Masai adult males.

to show they were, the genes of these 'marginals' would eventually have influenced populations inland.

Today, northeastern Africa is one of the most complex cultural mosaics to be found anywhere in the world. Layer upon layer of historical and prehistorical movement, expansion and decline are manifest in numerous languages and physical types. In the biased language of the times, many northeastern Africans used to be called 'Caucasoid-looking'. The simile should be inverted. Many Europeans and western Asians look northeast African, for the simple reason that the region is close to where they all came from. It is known that Sabaeans from the interior of Yemen migrated in numbers (particularly at times of famine) to Eritrea and Ethiopia but the main genetic influence in northeast Africa would have derived from within Africa.

It is to Africa that Europe and western Asia must look for their own origins. It is not just a matter of ultimate origins, such as those that link all peoples of the world as the children of an African Eve. African roots for western Eurasians are more recent than that. The populating of Europe by Modern humans was as recent as 45,000–32,000 BP and the further expansion into northern Europe a mere 15,000 years ago, so it is not altogether surprising to find a strong common resemblance on both

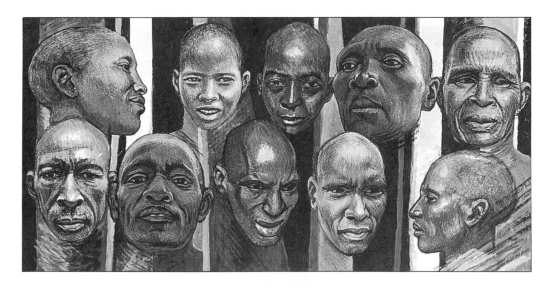

Eurasians. (top from left) West European middle-aged male (profile and frontal); German-Jewish youth; West European middle-aged female (frontal and profile). (below) North European elderly male; Central European middle-aged male; Western European adult male; Middle Eastern (Kurd) adult female; Pakistani adult male.

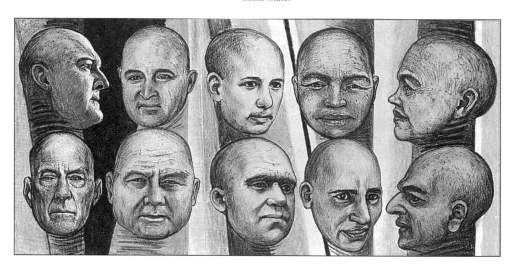

northern and southern shores of the Mediterranean. Of course, thousands of years of boat traffic, the ebb and flow of empires, immigrations and emigrations have left little chance of sorting out this stew of peoples. Nonetheless, it is clear from both fossil and archaeological sources that the contemporary inhabitants of the Mediterranean differ from their early predecessors principally in being slighter and less muscular; there have been few radical replacements or displacements other than the demise of Neanderthals in the north. What are called 'negroid' characteristics (those of tropical Africans) were very late to appear on the North African scene and although this is a widespread influence now, northerners were decisively different from tropical Africans until very recently.

Contemporary Arabs have not escaped the genetic mingling that has affected everyone else in the region. Again and again populations have traversed the Middle East, crisscrossing its narrows to Africa and Asia. From this land, which is both Africa and Asia, populations have pushed out at one time or another in every direction. The three most successful expansions were southeast into India, north into western Asia and northwest into Europe. As the genetic reservoir which spilled out into northern Eurasia, Arabians might be seen to provide an archetype for what is still commonly called the 'Caucasians'. In fact the Mediterraneans, Semitic-speaking Arabs and western Asians show no sharp break or

East Asians. (top from left) Siberian adult female; Yunnan adult female; Nepalese male youth; Vietnamese adult male; Siberian middle-aged female. (below) Alaskan (Esquimo) middle-aged male; Central Asian (Kazahk) adult male; South Chinese middle-aged male; North Chinese middle-aged male; Japanese elderly male.

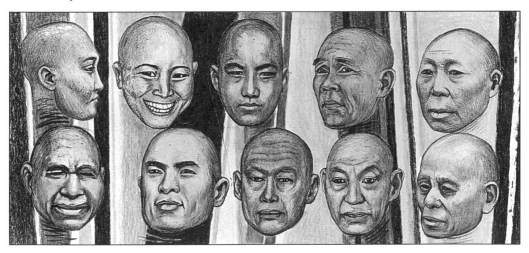

major genetic differences and they retain many anatomical similarities in spite of great individual variation.

The western Europeans are also fundamentally similar but have evolved the peculiarities of paler skin, hair and eyes (certainly within the last 38,000 years and probably in less than half that time). Within this vast region there are, nonetheless, some interesting gradients. The blood type 'r' graduates from a high incidence in the west to low in the east. Rh Negative (cde) has a very interesting pattern with very high frequencies (40–50+ per cent) in North Africa and western Europe, medium frequencies in tropical Africa and western Asia, graduating to near zero to the Far East and towards southern Africa. This looks like a North African gene that migrated with the colonists into Europe.

One very interesting pattern concerns skull proportions. Africans, western and northern Europeans, Arabs and other south Asians share in having relatively long heads, yet eastern Europe and central Asia share with the Mongolians (but not the south Chinese, Japanese and Eskimo) a high frequency (over 90 per cent) of short-headedness. This distribution has a remarkable coincidence with the empires and regions of conquest of the Mongol Khans. This broad belt of short heads corresponds to a zone where interactions between Mongols and Eurasians have been very extensive over many centuries, perhaps thousands of years.

Because the Mongolians showed some of the more extreme character-istics that distinguished eastern Asians from westerners, the term 'Mongoloid' has become the accepted term to describe all the paler, straight-haired, broad-faced and sometimes narrow-eyed east Asians and their Amerindian cousins. There is some justification for a common term, because the Mongoloids have a greater degree of genetic and anatomical resemblance than most people. Having more in common with one another seems to be due to a sudden and very extensive expansion at a relatively late date from one or more foci in the Chinese heartland. The Chinese, in particular, are perceived as peculiarly united because of their long history as a single empire or state. Their genetic relationship with the peoples of Indochina and central Asia is probably very complex, since the Chinese themselves probably arose from just one of several more or less discrete populations in eastern and southern Asia. Well-known genetic peculiarities are the possession of the 'Diego' blood type and the appearance of a bluish patch at the base of the spine in Mongoloid children (this may or may not persist into adulthood).

East of the Bay of Bengal, virtually all the mainland is inhabited by people of a Mongoloid physical type that seems to have expanded quite recently and spread very widely. However, some Indonesian and Indochinese populations probably combine genes from earlier colonists with those of this later dominant type. In Burma, the Mon are thought to represent the oldest residents, whereas both the Burmese and the Shan came down from the north. Thais too represent layers of superimposed cultures and peoples. Among possibly older types are the Mrabri, mountain people who were until recently foragers of wild foods.

A Chinese text, written 1,740 years ago, has been quoted describing the subjects of the Funan kingdom (today's Cambodia) as having black skins and frizzy hair. Although there are still relict groups of darker-skinned people in the mountains, this report, if correct, serves to underline how rapidly population change can take place. In any case, archaeological remains suggest that the Mongoloid assimilation or replacement of earlier Banda populations on the islands and tropical coasts of Southeast Asia was a very late development.

The densely forested Southeast Asian peninsula would have had a pre-agricultural population that occupied relatively narrow ecological belts within and around a much larger forested matrix that was mainly uninhabited. This population would have been relatively large and widespread by the standard of foraging economies and, having colonised all the healthy land suited to its subsistence techniques, would have arrived at some sort of demographic equilibrium.

This scene would have been eroded when the Mongols, in possession of a very intensive agricultural economy, moved in from the north to colonise the relatively empty valleys. Their economy supported much larger and denser populations. Because the effectiveness of its highly organised society would have depended upon numbers, it must have begun by concentrating in valleys that were well-suited to cultivation. From a number of steadily proliferating centres these nuclear populations then expanded in both numbers and territory, slowly squeezing out the foragers. Communities find an equilibrium only when the limits of finite territories and resources have been reached, and these limits are always subordinate to the current techniques used for sustaining that population.

There is little sign of such an equilibrium in contemporary Southeast Asia where the Indonesians are in a phase of explosive expansion. Indonesians (inasmuch as they can be lumped together) resemble other Southeast Asians in being predominantly derived from the recent southward expansion of Mongoloid people. They are more prone to wavy hair, have browner skins and a more indented root to the nose but they resemble other Southeast Asians in their blood group frequencies.

The general pattern in Indonesia is that Mongoloid resemblances are strongest close to the mainland, while Melanesian or Negrito features become more prominent to the east. However, the massive translocation of large Javanese communities into the western part of New Guinea (Irian Jaya) and to other islands of the Banda Sea is rapidly changing their demography. Until very recently the Makassar and Lombok straits constituted a major biogeographic boundary that drew a similar boundary between human blood groups.*

The rate at which prehistoric people dispersed has been the subject of several studies and calculations. Figures have ranged between 1 km a year for Neolithic farmers to 13 m per year for African foragers, with 0.1

---

*On the Melanesian side there were high frequencies of N, S, CDE and ADA 2. Indonesians west of Lombok, instead, share with other Mongols the 'Diego' frequency in the M group.

km a year for the first colonists of Australia and New Guinea. These rates have been computed on an almost geological time scale. The real rates of diffusion would have been by fits and starts and would have been slowed, speeded or even reversed by the state of the environment and by periodic pulses in technological invention and population growth. Thus people with a new food-getting technique or superior arms would have had little to check their expansion in territory or numbers when moving into thinly occupied lands.

The basic dynamics of foragers are determined by whether foods are concentrated or dispersed, reliable or unpredictable, static or mobile, and whether the type of food demands specialised hunting, transporting or processing.

For most situations the average number in a single, closely related band is about twenty-five individuals and the maximum number of bands that can keep in touch, exchange mates and have territorial agreements with one another is about fifty (although the average would be about half this). Thus the average 'tribe' would have comprised 300 to 1,000 people. When numbers dropped below this range there would have been the danger of being absorbed by stronger tribes (significantly the maintenance of a local dialect or language has been calculated to require a minimum of about 250 people).

In the dispersed conditions of a foraging existence, a tribe would fragment into two or more groupings well before reaching 2,000 people. Early accounts of foraging societies from many parts of the world attest to the hostility, suspicion and lack of contact that could arise between neighbouring tribes or clans. For example, the total number of languages in Australia before 1788 will never be known, but current estimates are in the region of 220. In the more crowded island of New Guinea there are known to be 1,000 languages.

A proliferation of languages and dialects would have been typical of all foraging societies. The range of densities would have been from about one person per 200 sq km in desert to one person per sq km. The highest densities could have been reached only in the best habitats, mainly tropical seashores and estuaries.

Analysis of gene patterns has suggested that long before the outer Pacific was peopled, some of the Polynesians' ancestors were marine fisherfolk sailing the South China Sea and living on its shores and many small islands (among which was Hong Kong). At this stage they would have been a peripheral part of an early Chinese gene-pool. The emergence of Polynesians and their expansion over great areas of the Pacific in the last 6,000 years is best explained by a complex interaction between people and technology. Initially Chinese fisherfolk spreading down along the Philippine coasts absorbed and synthesised techniques, mixing with the people they encountered in some places, avoiding them in others, winning here, losing there.

If genetic clocks are to be trusted (and it is as well to reserve some scepticism about dates), the Japanese geneticists K. Omoto, S. Horai and H. Matsumoto estimate that continental proto-Mongoloids arrived in the Philippines some time between 20,000 and 10,000 BP. They then posit further large-scale migrations of Austronesian slash-and-burn agriculturalists into the Philippines about 15,000 BP. These bouts of immigration were superimposed upon an earlier Banda population. The Philippines, therefore, got off to an early start as a genetic melting pot. The Philippines are also the most likely group of islands in which the Polynesians became a distinct South Pacific people. Here, they developed further the fishing and sailing skills that sent them surging over the Pacific about 6,000 years ago.

During this long spread of time the original Negrito populations were absorbed or retreated into enclaves. There are six distinct Negrito groups on six islands and at least eighteen population enclaves, most, but not all, are in mountains and it is interesting that originally seashore-based societies should have relinquished their prototypical habitat. It goes to show that marine resources are a prime focus for competition and that the most productive methods and the best gear wins. Nonetheless, every wave of newcomers would have shared in the genes of its predecessors and the Negritos themselves should not be thought of as the pristine aborigines of the country, for they too have been a part of the Philippines melting pot.

In their progress across the Pacific the Polynesians would have had many contacts with the Melanesians that were already established throughout the inner zone of South Pacific islands. In what is now called Micronesia essentially hybrid cultures grew up, as they did in those parts of New Guinea and the Solomons where Polynesian influence was not repelled. In the Solomon Islands detailed analysis by Jonathan Friedlander and his colleagues from Harvard has shown that several communities within this mosaic of tribes have had recent but strong infusions of external genes. In spite of this, there is little in the external appearance of Solomon Islanders to betray this heritage. All have very black skins and dense spiral hair, demonstrating that the genes for these features are very dominant. There is also the implication that natural selection must be very strong for intermediate skin and hair types to be so rare.

The Solomon Islands are a major interface between Melanesia and Polynesia. Just how the two peoples behaved towards each other in their initial (and also not-so-initial) encounters varied from total assimilation to outright war. One incident that is part of local history occurred off the north Solomon island of Buka and is reminiscent of the story told by the ethologist Konrad Lorenz (in *King Solomon's Ring*) of the fatal consequences of bringing two incompatible war strategies together. According to Lorenz, turkey cocks fight until one of them tires, whereupon the vanquished signals its subordination by stretching flat on the ground. Peacocks, by contrast, fight very fiercely with their spurs; as soon as the stronger bird proves his ascendancy its opponent must flee for its life. Should a turkey cock perchance fight and lose to a peacock, Lorenz laments that the turkey's inappropriate strategy ends in carnage.

The fishing village of Hanahan lies in a cove directly below the peak of Mount Bei, the 1,630 m volcano that dominates Buka Island. Bei's peak can be seen for over 100 km on a clear day and was a landmark for the Polynesian fisherfolk of Kilinailau Island 70 km northeast of Buka. By contrast, the existence of Kilinailau used to be unknown to the black Halia fisherfolk of Buka, because being a coral atoll it was beyond the horizon. When a single Buka fisherman was blown by a storm to Kilinailau, he was fed and entertained on the atoll, then sent home again. In the crowded and tribally fragmented Solomons it was normal to fight for scarce resources and raids were frequent; so the response of the Hanahan villagers to the discovery of their neighbours was entirely appropriate to their own traditions. The men took a fleet of canoes to the atoll, killed all the Kilinailau men and brought back the women to Buka.

The Halia fisherfolk speak an Austronesian language and previous incidents such as this may have been a major source of their Austronesian genes and language. Nonetheless, in their appearance they differ from the aboriginal Papuan-speakers of the Solomons' interior only by being taller and thinner. Raids like these had made the warriors of Buka feared all down the Solomons' coast, but with the coming of European colonisation and missionaries the raids stopped. In recent times Buka sailors regularly visit islands hundreds of kilometres away, including Kilinailau, where they are now on good terms with their neighbours.

A pattern that is repeated on several islands in the South Pacific is the presence of Melanesian-looking islanders speaking a Polynesian language. Thus the arrival of a few Polynesian immigrants on Ouvea, in the Loyalty Islands, led to the formation of two language groups, the Uea adopting Polynesian, the Iai maintaining the original Melanesian. No genetic differences have been found between the two communities.

The peopling of the Pacific is a story in itself, but here again genes can provide important insights. One such interesting detail of the Polynesian diaspora concerns one of the many genetic bottlenecks these people must have gone through as tiny groups discovered distant islands. The two most remote areas, Easter Island and New Zealand, were apparently both colonised from the same source (probably the Marquesas). In all three of these very scattered islands two antigens have been lost (HLA-B13 and B27).

Genetics sometimes comes up with some unexpected results. For example, of all Asian and Pacific peoples, the Amerindians have been found to be closest to the Japanese on the basis of genetic markers in their blood. This is less surprising when it is remembered that the Amerindians derive virtually all their ancestry from the northeast Asians between 22,000 and 11,000 years ago and that many demographic changes have taken place in northeast Asia since then.

It is interesting that the Amerindians should be less related to modern mainland Asians, because this supports other more general evidence that the modern dominant Chinese population derives from recent inhabitants of the Han and Yangtse Valleys. Starting late from a south-westerly centre of expansion in Korea, Japan was invaded by unequivocally Mongol people (the Yayoi as late as 2,300 years ago). The genetic links between America and Japan are best explained by the Japanese retaining more of the Jōmon and northeast Asian inheritance than mainlanders.

The theory that the colonisation of America is a recent event and that all Amerindians are closely related has some confirmation from a recent wide-ranging study of teeth. The dental specialist C. G. Turner has found less variation among Amerindians from both continents than in two relict groups in Alaska. As old residents of prehistoric America's port of entry, Alaska, Eskimos and Aleuts seem to have kept more of their Old World variation.

The expansion of Mongols all over the Pacific and to the western side of the Pacific Ocean is well known. Their movements westward across the Indian Ocean tends to be forgotten. Its outcome too was very different. In spite of being much more recent than the main concerns of this book, it is a tale worth telling, partly because it demonstrates that the marine migration routes used by the Banda were followed by much later migrants.

While favourable winds allow quite small sailing craft to travel back and forth along south Asian shores and along the entire length of the East African coast, we have become so used to the political, bureaucratic and nationalistic barriers that now obstruct such enterprises that it may seem much more difficult than was actually the case. In AD 1235 Ibn al-Mujawir described *outrigger canoes* sailing from Madagascar to Aden and back, merely running with favourable winds and following the coast.

A flourishing trade route was developed between Indonesia and East Africa about AD 300, which reached its peak between AD 700 and 1000. Hundreds of boats travelled back and forth and the powerful princes of Sumatra and Java used African produce to pay tribute to their own overlords, the emperors of China. Thus, African commodities and slaves found their way from Africa to Indonesia. Even a giraffe from the Horn of Africa ended up at the Chinese court, where it was taken to be the mythical kirin, a creature of good omen. Although fairly substantial ships must have been used to carry such cargo, most of the boats that conducted the trade were little bigger than dhows, but probably made up for their lack of size by sheer numbers. In AD 945 an armada of Sumatran ships attacked Kanbalu (probably Zanzibar) to secure it as a trading port.

So for more than 700 years quite small boats were confidently used to make the trips back and forth between East Africa and Indonesia. Had the boats hugged the shores, a one-way trip would have been over 12,000 km, but the distribution of Indonesian styles of boat construction shows that their journey island-hopped from Sumatra to Sri Lanka, the Maldives and hence to the East African coast, possibly via Socotra. This 'short cut' would have been about 9,000 km, much of it in dangerous open seas. Yet such was the Indonesians' confidence in their skills and craft, and such were the incentives of trade, that these long journeys were undertaken as a matter of course. With good winds a single journey would have taken about two months.

The Indian Ocean coasts were an easy diffusion route for simple types of innovation in boating or fishing techniques and discovery of the trade route and its potential commodities was probably the result of much earlier small-boat exploration that went closer inshore. The *Periplus of the Erythrean Sea*, a travel book written about AD 100 mentions trade in coconut oil and the presence of sewn boats along the East African coast. Both must have had eastern sources.

However, the monopoly of trade by the princes of Sumatra and Java, during the seventh to the eleventh centuries, was more than a mere diffusion agency: it was effectively a maritime empire steadily built up over a very long period. Large numbers of Southeast Asians would have settled on the littoral of what is now Tanzania and they probably pioneered the trade routes to the interior later to be seized by the Shirazis and Arabians. Even after the takeover, a twelfth-century Arab writer, al Idrisi, described Africans speaking the Indonesian language and continuing to trade with both Java and Madagascar.

One justification for this incursion into recent history is its illustration of differing fates for the same people in neighbouring areas.

There can be no doubt that the primary attraction of the western Indian Ocean for the Indonesians was highly valued commodities from Africa: metals, ivory, gums, lacquers, beeswax, incense, ambergris and slaves. In the course of setting up this coastal empire, the Indonesian boats encountered Madagascar (probably about AD 440 although the earliest dated remains are AD 600). This island was uninhabited, so any labour required for the extraction of such produce as it offered had to be imported (extinction of six species of giant elephant birds by Indonesian pioneers is some indication that the island had never been settled before). Madagascar would at first have been no more than a colonial satellite to the African mainland, where most of the Indonesians would have been settled. They would have lived in houses made of perishable materials, probably of the type found in Indonesia today.

At the north end of Lake Malawi, the Nyakyusa people are unique in Africa for their Southeast Asian style of building bamboo and thatch houses. Indonesian-style canoes with the peculiarly Asian quirk of a bifid prow still sail on Lake Victoria. Sewn plank canoes, called *Mtepe*, are also common on the coast and the fisherfolk have continued the Southeast Asian custom of painting or incising an eye on the prow. Their favourite musical instrument is the Indonesian bar-zither and they shred coconuts with a stool scraper which originated in the East.

In the thirteenth century Ibn Said wrote that Malagasys had penetrated into the African interior and gave their name for Madagascar, 'Gazira al Komr' or Island of the Moon, to a mountain in the African interior. (If this is not mere fable-making, Kilimanjaro in Tanzania is a more likely candidate for their legendary 'Mountain of the Moon' than Ruwenzori, which has now acquired that name.)

In central Tanzania there is a small tribe of rather distinctive-looking people, the Turu, who have the oral tradition that their ancestors came across the sea from an island in the east and then followed a great river into the interior. Although several observers have claimed that the Turu 'look Indonesian' they would undoubtedly be regarded as African by the unbiased eye.

These then are the rather inconclusive vestiges of an empire that lasted at least 700 years on the African coast and probably exerted some influence over much of the eastern interior during this period. By contrast with their almost total eclipse in Africa, Indonesians became the dominant population of Madagascar. At the same time as the Arabs began to usurp their role, the colonists' links with Java and Sumatra were getting more tenuous and Madagascar became the obvious refuge for displaced Indonesians from the mainland. The Madagascar channel is only 500 km wide but its winds and currents had been sufficient to deter previous colonisation by Africans. Whether an exodus of 'expatriates' from Africa helped swell numbers on Madagascar is not known but the huge empty island permitted a rapid growth of population from a relatively small founding group. Although there is a high proportion of Malagasies who look Afro-Indonesian and the vast majority are likely to have mixed genes, there has been a significant separation between brown straight-haired and black spiral-haired populations. The former dominate all the uplands of the interior while the latter occupy the coast and lowlands. It is possible that a demographic pattern similar to that of the Caribbean Mosquito coast is involved.

In this brief world survey of what are commonly regarded as 'the races' I have tried to stress that the differences that can be seen in modern people speak of four more or less interlocking processes that can be illustrated by living people. The first is simply the product of time; genes mutate and people change or evolve as time passes. The second process concerns space; isolation (partial or complete, temporary or prolonged) permits genes to change in ways that become specific to one area and different in another. The third process is adaptation to particular climates, habitats and to the physical or physiological demands of specialised ways of earning a living. Finally, there is the greatly underestimated influence of gene recombinations or interbreeding. The neglect of this subject by scientists and the reluctance of many ordinary people to contemplate this central facet of human identity is a major theme of the next chapter. Eve's children have been, are and most surely will be hybrid.

8: A Family with Baggage

At the age of about fourteen I had to find my way from a school in Europe to my home in Tukuyu, East Africa. The first leg of the journey was by flying boat to Lake Naivasha in Kenya. There the passengers had to fill in immigration forms where one question was 'race?' It is perhaps a juvenile instinct to dislike being categorised, so I wrote 'human'. After our passports and forms had been collected together and taken to the office, the clerk reappeared to say the immigration officer wanted to see me. 'What's this nonsense, laddie? As far as this form is concerned you are Caucasian'; and 'Don't be cocky' when I mumbled something about being neither a cork nor an Asian.

My gesture may have been absurd but there are also frequent absurdities in the taxonomy that is so solemnly put down on forms and computer tape. Earnest attempts to determine when Mongoloids, Caucasoids and Negroids diverged from one another are the object of sophisticated technology, long hours and pages of tedious discussion in learned papers. The mismatch here is between the application of supercomputers and the results of genetic wizardry to what has been called folk taxonomy. Apart from the difficulties of defining these supposed taxons, it is more important to know *how* any sub-grouping of humans came into existence, *where* and *why*.

The greatest interest of parts and the comparison of parts is that they help define the whole. What it is to be a human will always be more complex, interesting and important than what it is to be any particular type of human.

The reluctance to project any of the demographic processes we take for granted in the present back into a primitive and 'biological' past seems to have had several causes. Firstly, an infatuation with contemporary hi-tech tends to underrate prehistoric technology, mobility and intelligence. Secondly, ethnocentricity is so ingrained and global consciousness so new that the prehistory of tropical people (even when known) has been considered too marginal to deserve serious consideration. Thirdly, there has been too close and easy an identification between the geography of man and the Ethiopian, Oriental and Australasian biogeographical realms. Fourthly, the allocation of racial 'types' to such continental and island regions has reinforced types of race consciousness that have been grotesquely rigid, at odds with the facts and a real obstacle to understanding and reconciliation.

Moderns have radiated very recently; our differences are open to enquiry and our diversity is a matter for celebration, not suspicion.

The definitions that arise from mapping the human genome will eventually totally restructure our perceptions of what we are. It will be as useful to medicine and self-knowledge as the discoveries of anatomy and physiology were during the last century, but the need to start redefining what it is to be human cannot wait on tomorrow's cartographers.

Adam and Eve have an enduring appeal and 'the sons of Adam' is still a current synonym for 'mankind'. These two patriarchal terms are already challenged by scientific discoveries about the very nature of genealogy. Only 'Eve' and her female progeny offer us our most ancient line of descent (through mt DNA). When male-influenced nuclear DNA is analysed, different and more confusing trees appear because of all the recombinations that are incorporated.

More cryptically, the innocent 'Bin Adam' contains within it an assumption that we define our humanity through an affirmation of our ancestry. The reality of ancestry, something deeply felt in every family, has in the biblical tradition been passed through a filter of belief. Ancient texts and everyday speech affirm that we *believe* in our ancestors, we do not *know* them. Now, for the first time, knowledge begins to be a credible and preferable option to belief.

There is yet another dimension to having new definitions in the air. Mythic or semi-mythic ancestors of all sorts are not mere religious conventions, picturesque metaphors or heroes of history, they have become embedded within the structures of speech and identity. An isolated, unencumbered individual, without any social or other baggage, would find it difficult to be taken seriously if she or he claimed direct descent from Adam, Shem, Ham, Abraham, Judah, Ismail, Sheba or even Owen Tewdwr. Yet all are real of supposed ancestors that validate contemporary names, identities, institutions and ethnic or religious groups. All, therefore, have to be taken seriously as part of the baggage members of the human family inherit.

295

The language groups called Semitic and Hamitic have become acceptable scientific terms, even though they are rooted in Mesopotamian myth (Shem and Ham were Noah's two sons) and the mistaken idea that these terms describe real 'races' is still believed. Judaic and Ismaili institutions are built on the authority of their respective ancestries. Two contemporary royal lineages trace their supremacy to a Sabaean queen and a Welsh chieftain.

For an increasing majority of people there is impatience with identities, identifications, institutions and entities that are described and validated through the shallow myths of a recent but already well-forgotten past. The majority of people have rather few remembered ancestors and those they have were so thoroughly uprooted and disinherited that traditional reverence for ancestors is more and more a dead duck. For the majority, identity is imposed by accidents of birth and geography. Most allegiances are to Religion, Nation, State, Flag, sometimes to a class or trade union. Increasingly, common cultures are definable by the personalities and programmes on television or radio. In these, as in older cultures, identities are as much imposed by others as inherited or self-made. You are no longer simply a butcher, a baker or a candlestick-maker, but you live up what it is to be a yuppy stockbroker, a hip-hop rapper, a biker, a mercenary technician, an ivory tower academic or a blue-collar miner.

A big difference between modern and earlier identities is the ease with which they are shed. Change your job, address, passport or spouse and you move into what is commonly called 'another world'. This is less easy if you are very poor or very oppressed. For these, the majority, jobs are on a take-it-or-leave-it basis and choices are limited, especially under authoritarian regimes and in 'command economies'. What jobs there are become state-ordered condescensions or rewards for loyalty, and the only real alternative is emigration. As a result, massive emigration has been, is and will continue to be an integral part of life.

Those who emigrate or change their jobs often believe they can become a new person. In fact, the paraphernalia for that new person is usually already in place and monitored by adopted peers who are also already in place. It is very difficult to shed baggage without acquiring another secondhand set.

An important example of an entire community rejecting a 'given' identity and firmly asserting their own self-definition instead is the civil-rights movement in the USA. It is often assumed that this was a peculiarly American affair, most especially endemic to the medieval Southern states. Although this movement has its roots in a peculiarly American history and an incomplete emancipation from slavery, the leaders of the Civil Rights cause voiced a much more universal demand. This was their refusal to occupy tamely a niche that had been ordained for them in an uncompromising and totally oppressive past. Since those who had manufactured this intolerable system had chosen to emphasise their own distinctness as 'white' people, the oppressed, naturally, found their common identity in asserting the opposite: 'We are the *black* people.' This was more a reaction to America's history than an expression of real racial homogeneity. Where the pre-revolutionary jazz trumpeter Louis Armstrong had sung an ironic lament with a humour born of acceptance – 'Why did I have to be so black and blue?' – black Americans of the 1960s and 1970s proudly raised their fists and asserted their own physical vitality and political aesthetic – 'black is beautiful'.

This very American revolution became internationalised in a well-publicised and picturesque way. At the 1968 Olympics in Mexico City two American medallists, Tommie Smith and John Carlos, stood tall, handsome and triumphant on the podium and gave the 'Black Power' salute. While accepting the medals for the United States of America, these sportsmen used that simplest of gestures, a raised arm, to assert a second loyalty and additional identity.

In a less flexible and idealistic society, America's civil war and civil rights revolution might have fragmented the country along racial lines (as South African apartheid governments attempted with the ill-fated Bantustans). Instead America is a society that has sought to be both democratic and nonracial in its principles (even if it falls far short in practice). Narrow ethnic chauvinisms, each reacting separately to a bitter past, a disillusioned present or an anxious future, seek to discredit and reverse all that has been achieved. The comings and goings of sects and movements is nothing novel and the 'tribes' of modern America are so newfangled, contrived and synthetic that few of their tribesmen and women are likely to have the convictions to sustain their allegiance for anything that might approach a lifetime. The ultimate strength of America ought to lie in an enduring defence of justice and opportunity.

It is a society that has been rich enough to be generous and confident enough to be tolerant. Should the decline of America's wealth and the shaking of its confidence ever lead to its social disintegration, it will be more than America's failure. Under the most auspicious of circumstances humanity will have failed to recognise that it is a family.

Without very vigorous efforts to combat it, racial prejudice is either a cancerous growth or a formally accepted aspect of life, as it has been for millennia in racially diverse but caste-ridden India. Here, the largest share of resources has always gone to higher castes, while the lower castes were kept from anything that might make them independent in mind, choice or action because of the landowners' need for compliant labour to work their lands and factories.

Every external effort to empower or enrich the lower castes has been aborted by the sheer power and inertia of a deeply ingrained tradition. Only in the cities and abroad has the stranglehold of caste been loosened. From an egalitarian point of view this is almost the worst outcome, because prejudice has become self-perpetuating. A diversity that might otherwise have broken down into a more and more mixed society is merely reinforced by absolute barriers to marrying (or even dalliance) between members of different castes. The differences between castes are further reinforced by absolute partitioning of jobs, with the lowest castes doing the least pleasant or most 'polluting' jobs such as tanning, waste removal and butchery.

When two visibly different communities live together in the same society, there have usually been historic events (and sometimes special skills and attitudes) that lead to a similar partial or almost total demarcation of jobs. That partitioning has within it the potential of becoming a *de facto* caste system if the two communities retreat into separate suburban castle and city ghetto, Brahmin bungalow or Harijan slum.

This sort of separation between two different communities is always accompanied by the creation of a third community, which accommodates those who belong to neither of the main groups. The natural dynamic is for the highly diverse third group to become the majority eventually. In South Africa, that outcome was artificially slowed by physical separation, social pressure and legislation to punish what was called miscegenation. Until very recently South Africa was a museum of mankind's most primitive sectarian instincts and fears. In their belief in 'blood' and 'race' nationalist Boers pursued a Machiavellian protection of their own interests and maintained stolid attachment to a pseudo-zoological classification of humans that rendered anyone outside their system a 'hybrid'.

There is an ancient and deep-seated distaste for and unease about the hybrid. Deriving from a Roman epithet for a domestic sow that mates with a wild boar, the word was transferred to describe the offspring of Roman citizens who married foreigners. When social ranks are broken or racial fences crossed, new, unclassifiable children are born that threaten classification itself. Class, race, caste and pedigrees are among the institutions that give stability to society, so the depth of hostility to the hybrid is manifested in numerous derogatory words and phrases – bastard, mongrel, crossbreed, half-caste, *mengen*, fifty-fifty, quadroon, mulatto and mule. Countries in which such expressions of prejudice are flaunted openly are fortunately getting rarer but their presence, open or concealed, is a sure sign that racism is a reality and is still alive in that society. For a child born in the no-man's-land of a racially polarised society, for the victims of pedigree bigots, there can only be a profound and unsettling anger. In South Africa, 'coloured people' are among the most thoughtful, passionate and articulate opponents of the gross and primitive Babylon that has so systematically humiliated them. In other less medieval but still racially polarised countries it is the person who is of 'neither, yet both' communities who is likely to be the most radical in calling for social or political change.

An alternative course is to enter the most intensely competitive of the tournaments and arenas where racial prejudice wilts before excellence of performance. It is no accident that many great public performers in the arts, music, dance and sports are the children of ethnic mixtures, minorities or ghettos. They have the incentive to prove to others that they are as good as or, preferably, better than anyone else. Over this century, in the boxing rings of New York, a succession of different ghetto boys have won the belts and medals – Irish, Italian, Jewish, Hispanic and Negro. The last group have had the strongest incentives to excel, and competitive sports have become a special arena for black minorities in many countries. In the Olympic Games no one can deny the disproportionate number of winners, especially in track events and boxing, that are dark-skinned and of at least part-African origin.

If a large part of the explanation for this lies in the high motivation of immigrant, minority and 'mixed race' communities to prove themselves in the highly competitive Western cultures, this cannot be the whole story. When schools and colleges, the army, police and municipalities in Kenya and Tanzania began to organise sports in a systematic way, their best athletes were quick to win gold medals in Olympic track events. Few would argue with the statement that black men and women are outstanding runners, strong and resilient athletes with exceptional speed and stamina. If the blistering rays of a tropical sun bouncing off waters and beaches of the Banda Sea once weeded out countless numbers of children and adults, there is a certain justice that their most distant descendants should vindicate the cost of that selection with medals.

There is, of course, a profound reluctance to believe in genetic advantages. Taken to its extreme, the persistent triumph of track athletes with a Banda inheritance would discourage others from competing. Fortunately, the wide range of aptitudes that are exercised and tested in Olympic sports tend to even things out with long-distance swimming and other medals more widely dispensed to individuals with especially good lungs, hearts, liver and buoyancy. The disbelief that genes can confer advantages is shared by the public at large and many scientists; but, as soon as this argument is seriously developed, the ground shifts from biology and genetics to ideology.

Quite properly, the opponents of genetic advantages recognise that very dangerous political forces have been unleashed by wholly spurious ideas of racial superiority. The horrors of fascism are still so close that a tacit moratorium has been declared on serious discussions of race for the last fifty years.

Of course racial prejudice is nothing new and sophisticates have always despised those that have yet to acquire their own mastery of technology or political power. In 77 BC the essayist Marcus Cicero of Arpinium wrote 'Do not obtain your slaves from Britain because they are so stupid and utterly incapable of being taught that they are not fit to form part of the household of Athens.'

When I first drafted this book I had thought that enough time had passed for both the immorality of racism to face no serious challenges and its stupidity to be wide open to ridicule. This assumption has been overtaken by events as the racial impulses of some Europeans have reasserted themselves. While the roots of this resurgent xenophobia may lie in the stresses of political change and social insecurity it can still draw on an obsolete literature that includes long academic treatises linking race and intelligence.

Books on race and human diversity often used to discuss intelligence as if it were a projection of school. Teacher awarded grades to rows of cardboard pupils; the latter were scored with point systems by the writer or his illuminati. IQ quizzes and various types of intelligence tests were invoked to give a scientific gloss to a classroom exercise that verged on comedy. Less comic was the application of such ideas to 'races' and their capabilities and the willingness of readers to be taken in.

The inventor of the first intelligence tests, in 1905, was the French psychologist Alfred Binet, who devised them as relatively simple scholastic tests in order to help pupils with learning difficulties. Binet had a name for those who seized upon his functional scholastic tests as if they were the final measure of inherent intellectual quality; he called them 'brutal pessimists'. If the protean and liquid form of human intelligence and imagination is ever to be measured and contained, it will not be with the paraphernalia of simpletons. Least of all can it be touched with the brutal pessimism and misanthropy that underlies racist attitudes.

There has been far too little serious challenge to the concept of race itself. All too few people have taken on board the idea that the dominant 'markers' for race, such as blackness in Africa and whiteness in Europe, represent small but useful adaptive packages that date back to a distant prehistoric past. In both instances they conceal a more significant and interesting heterogeneity within those continents. These markers are minuscule particles of protean wholes. It is rather like the mistake of early taxonomists, who thought that the black and white morphs of reef herons were different species. The so-called racial markers signify no more than a wholly natural mosaic of strengths and weaknesses. Human diversity is a festival parade, not the primitive source of suspicion, fear and domination it so often has been. Diversity is an inescapable part of our biological heritage: it will help protect Kumar from one thing and put Kanesia at risk from another; it is both a source of security and a liability. Yet an awareness of our weaknesses is something that technology and medicine have encouraged us to forget. The media have seized upon our fantasies of invulnerability, so we have the absurdities of Superman, Rambo and Batman. Flights of the imagination, perhaps, childish, yes, but pathetic when that kind of fantasy bolsters an adult ego with supertoys and superweapons.

Of course they're only synthetic modern versions of old heroes but these macho concoctions are relevant to my larger theme in that the possession of lethal technology is central to modern myth-making and escapism. The hero is swept from one adventure to another by pressing the keyboards on the spacecraft's console or the trigger on a laser ray gun. The key to the viewer's escape into make-believe (TV dials and buttons) mimics exactly the technology employed by the hero.

The ways in which members of an individualistic urban society vent or express frustration at their condition is very different from those of more socialised communities. For example, there is a very old tradition in the Mediterranean that has been seized upon as a vehicle of expression by formerly colonised people who were at least not wholly European. The pre-Lenten binge that used to be called *carne vale* or Carnival may well have had pagan origins. It is essentially a Latin festival formerly celebrated mainly by Portuguese, Spaniards and Italians.

In the New World, and most especially in Brazil, Venezuela, Mexico and other Latin American countries, the Carnival became the one event of the year in which the poor, the despised and oppressed were free to shine, sing and dance in a public parade where all the constraints of a cruel and exploitative society were temporarily suspended.

Small carnivals flourished for centuries all over South and Central America. In the melting pots of Brazil and the Caribbean they really took off and captured the attention of a much larger audience. In the cities of Rio de Janeiro, Port of Spain and Kingston carnivals enlarged and took on the big-city look, with more bands competing, more and bigger companies and floats. Exported to Notting Hill in London and the streets of New York, the carnival became truly metropolitan in character while remaining a Caribbean initiative. In spite of all the glitz and the electronics, the raw materials of carnivals are the human body, some touches of colour and things that will make some sweet noise. It is still a festival where the poor are, for a moment, in charge so that carnival is also an affirmation of community and self-sufficiency.

It is not just the scale of the carnival that is enlarging, its constituency is growing and, as the people who participate become more numerous and diverse, this ancient traditional festival will continue its protean course, it will change because its essence is change – transformation – it is affirmation in the face of denial. It is a symbolic expression of humanity. It is nakedness dressed up.

If carnival has an element of escape into fantasy, one of the realities from which the urban poor seek relief is the inherent anxiety and precariousness of people without land and crops to fall back on. In this, urban folk differ from almost all pre-industrial societies but they do share one option. There is always the more literal escape from the intolerable, which is *to move*. Migration is what took carnival to the New World and back to the Old. Migration has shaped the modern world, will spring many surprises in the future and, as previous chapters asserted, has a venerable past.

Can there be any relevance in the movements and population patterns of prehistoric foragers for the present churning-up of billions of contemporary people? I think there is. For a start, the contemporary patterns of genetic mixing are no more than a dozen or so generations deep. Prehistoric patterns involve much smaller numbers of individuals but an inheritance of thousands of generations. Since genes change rather slowly if they are not augmented by migrant genes, is all this recombination something new? Genes had a much greater mobility in prehistoric people than has generally been allowed for. Although there is much to support the idea of prolonged genetic isolation in Australia, Hokkaido and a few other enclaves, the major continental populations cannot and should not be conceived as monolithic races. The erection of 'racial types', whether crude populist stereotypes or more subtle and earnest models constructed from statistical frequencies, is wrong. They are not only likely to be false as scientific models but also wrong at the popular and moral level in that they create false entities and false identities too.

When the four or five classic 'races' are assumed to lack the complications of a mixed genetic past this is merely a convenient convention. It makes for a much tidier model of humanity but I am convinced it is almost wholly false. While genetic studies at their present stage of development can neither confirm nor deny my broader hypothesis, they are consistent with a great deal of complex interconnecting and peculiar interlocking of patterns. They are also consistent with the proposition that small clutches of dominant and adaptively favourable genes can give a superficial similarity to people who are actually of very diverse origins. There are alternative explanations for such diversity but migration, mobility, intermingling is my bet. My principal example has centred on Africa and the Africans because they are genetically the most diverse people on earth, yet the most prone to be lumped under that single, simplistic category – blacks.

It is said that the wife of a senior cleric, on hearing of Darwin's ideas, exclaimed, 'Descended from the apes! Let us hope that it is not true, but if it is, let us pray that it will not become generally known.' A similar fear exists among racists, that they might be found out to be as mongrel as anyone else, so it is important that our mixed ancestries become generally known.

It should be more generally known that Europeans are mostly aberrant and relatively recent migrants out of Africa and the Middle East, with later infusions from south and east; that India and Southeast Asia have been even more thoroughly churned; that the Japanese are mainly a mix of Koreans and Ainu; and that the great majority of people were already thoroughly mixed before this millennium began. The cultural traditions on which nationalists and racists like to set so much store are seldom any more than fragile, mutable and recent expressions of a way of making a living. As the pace of redundancy for subsistence techniques speeds up, it only serves to emphasise the flimsy and transitional nature of all cultures.

In the meantime, migration and genetic mixing will accelerate under all sorts of unpredictable forces. More and more people continue to flee from the terrible technological and administrative power that has got into the hands of some of their compatriots. They seek hope instead of the hopelessness inflicted by greedy and incompetent politicians. They seek refuge from the criminal lusts and ambitions of military rulers. They seek relief from the hypocrisy and mental slavery imposed by religious bigots. They seek food when one or more of these groups have so hijacked the state that even the means of sustenance are denied them. All these forces are depersonalised and sanitised into abstractions – oppression, fundamentalism, corruption, war and famine – and their victims are meanwhile reduced to refugees and beggars.

The real events, as opposed to the abstractions, have names, dates and addresses. It would be encouraging to believe that the world will eventually change its texture under the influence of beneficent and liberal ideas and policies. Sadly, it shows few signs of doing so, and the main incentive to get up and go will continue to be sheer desperation. Fifty years ago the obscenities of Hitler's Nazism drove millions of Jews into extermination camps and the survivors into hapless Palestine. The Americas have absorbed all the discontented of Europe, the captives of Africa, the entrepreneurs of Asia and, to a greater or lesser degree, jumbled them all up with the previous Amerindian colonists. That all began with a queen's appetite for gold in 1492.

Two hundred years ago the corrupt monarchy of a naval power found it had the means to send its potential revolutionaries, critics and all manner of other social and criminal rebels to the other end of the earth, to Australia. That became the second major phase of colonisation for an entire continent. The decline of another naval empire in the twelfth century left its Indonesian colonists stranded in Madagascar, where they too founded a new nation.

Such simplistic vignettes do no justice to history but at least serve to emphasise that opportunism is an enduring trait in human affairs. The genetic impact of future movements is unlikely to match those of the examples just cited, in spite of transport having moved on from canoes and horses to the chartered jets of today's refugees and migrants. It is not only the dispossessed and the pawns of military and imperial players that are shaping the texture and variety of today's world. Tourists and students vie for cut-price fares, commercials and technicians fly the world in pursuit of business and these too are potent initiatives that bring different people together. In the process most societies are becoming more multicultural. All this is a direct product of technology and innate human opportunism. However, prejudice remains a part of tribal existence everywhere and is a common response to anything that makes individuals different, from malformations to skin colour.

If ignorance is the mother of prejudice it is perhaps no accident that the United States, a country founded upon high democratic ideals and the pursuit of knowledge, was the first to articulate the ambition to become a truly nonracial state. The ex-imperial nations of Europe, notably France, Spain and Britain, have attempted to follow similar principles and policies. Like America, they have had to convert a blatantly racial, unfair and domineering past into a nonracial and fairer present. Neo-fascist, tribal, nationalist political groups continue to hamper their success and prejudice based on the crudest of signals – colour and dress – is still pervasive and pernicious.

While all the main world religions acknowledge the unity and brother-hood of man and cultivate moral and spiritual values that ought to reconcile conflicts, they are not always successful. Followers polarise by sect, religion or nation and confrontations with nonbelievers can license hatreds that can be close to or effectively racial. Indeed, all international conflicts that align along geographic and racial boundaries have the potential to be race wars. Even within countries, the struggle to get or to keep the largest share of the cake can lead to ugly sectarian or racial confrontations; South Africa, Fiji and Sri Lanka are modern examples of how readily people align on racial lines.

Many of these are battles for resources in which the contestants remain crudely tribal. Traditional territories are a large bone of conten-tion. As resources get scarcer and populations get bigger such territorial conflicts can be predicted to become more intense and the sharpest conflicts may well remain *within* rather than between countries, in spite of the trend towards ever greater genetic mixing in almost all major societies.

This is not to say that the world will become homogenised. Even assuming increasing levels of mixing, distinct regional differences would remain both within countries and most markedly between continents. Recriminations between the descendants of different stocks will con-tinue to draw on historical injustices to fuel antagonisms within many societies. The many minorities that feel their identity has been destroyed or swamped will seek to find new allies. Through improved communi-cations and travel, local minorities will enlist with larger groupings of dispossessed and their sympathisers. New alliances will create new and surprising fellowships and new political leverage. After enduring local impotence for generations many minorities are finding a new potency in the larger world outside. It is in becoming *less* isolated, not more, that they will find their renewal.

History is no provider of precedents for present morality. Nor can history provide very strong moral cases for 'possession of the land' by any racial group. The dominant population in most large nations has displaced earlier, smaller and usually somewhat different populations. In every case large numbers of animals and plants have been exterminated along the way. In every case the means of defeating, swamping or absorbing the previous population can be traced to some technical or organisational complex that has permitted the invaders to outwit, out-populate, outflank or out-shoot the natives. If a new morality is to emerge it must come through diminishing, not increasing, the distances between us and from a greater determination that technology will be used less to take advantage of someone else than to offer knowledge where now there is ignorance.

If the main trend in the biggest and most influential countries is towards nonracial societies, what will be the long-term genetic trends (over centuries rather than decades)? The most obvious outcome will be a decline in recessive genes and an increase in dominant ones. This will certainly have a long-term effect on complexion. The genetic dominance of dark eyes, skin and hair is likely to change current blond/non-blond ratios. Whatever physiological or other advantages might have operated within its small and semi-isolated place of origin, blondness came to be associated with some technological and cultural traits that played an empowering role at certain critical periods of history. These attributes coincided with technological inventiveness and social dominance during the brief excursions of imperial Europe. This assisted a spread of blonds that would be surprising were it not already so familiar. A late 'sideline' has spread very widely all over the world and achieved quite dispropor-tionate political power. Now that their many technological inventions have begun to be more widely shared and Western monopolies of control over world resources have begun to be eroded, the numbers of blond people and their genetic influence can be predicted to wane as the ratio of recessive genes declines in proportion to main-line dominant genes. Nonetheless, blonds will remain a sizeable and conspicuous minority in many countries.

What is certain is that people will not become more alike but rather more diverse as more and *new* recombinations appear. The human family ought to get not less but more interesting. Whether it can remain healthy, happy and wise enough to enjoy being a family will depend upon the ends to which its techniques and technologies are turned and what is done to the earth that sustains us all.

9: The Sorcerer's Apprentice

On a global scale there is a continuous history of cultural development that links prehistoric hunters with today. The late discovery of Stone Age cultures in New Guinea and ancient fishing societies throughout the Pacific is a reminder of how close we have been to our cultural and ecological roots. It is frightening too how quickly loggers and fishing fleets with facilities for mass-production not only diverted these resources to distant metropolitan societies but stripped and degraded them in the process.

It has been intrinsic to history that each major development in food production or technology displaced and invalidated what went before. The fact that one system triumphed over another was enough to make its superiority self-evident, or so it seemed to the triumphant pioneers (and eventually to their demoralised predecessors, if they survived). When a new system feeds more mouths and breeds more babies, its children quickly become the majority and majority views are the dominant ones. During the last century the most pervasive and least questioned idea to take root all over the world was the belief that humanity will find some form of salvation through technology.

When visitors from the industrial world first descended out of the blue upon pre-industrial people, the strange traders and sailors were usually of less interest than their paraphernalia. In some particularly remote areas such as the South Pacific, trade goods became the object of cargo cults, in which it was believed that supernatural forces or gods had decided to send cargo for which the all-too-human Westerners were mere couriers. How could they guess the future price to be paid by their children for their axes, weapons and bulldozers? How could they envisage the price already paid by grimy workers in distant manufacturing towns? The inhabitants of industrial countries share forms of cargo cult in their belief in rights to the fruits of technology with no conception whatever of the costs.

The material origins of all modern technology and the vast losses of energy involved in its manufacture are easily forgotten. Since the pace of invention has never slackened, most people still envisage technology as a progressive force that has gone from strength to strength. This step-by-step progress can also be projected back into the pre-industrial and pre-agricultural world. Among the countless losers in history and prehistory were many who (precisely because they were not the winners) lamented the loss and the waste that went with each innovation in the exploitation of nature.

As cultures came and went, members of declining or defeated minorities, forced to live through the complexities of a marginal existence, would have been better able to see through the simplistic certainties of those who were in the ascendant, though they could not escape their effects. Sometimes, like the Amerindian hunters, they could only stand in neutered horror as they listened to steel axes and saws carving up the cadavers of their ancestral woods and watched their clear waters fill with the gungy run-off from new fields and clearings. Over many centuries the complaints of the dispossessed have been drowned out by the clamour and claims of new proprietors. That is as much a contemporary reality as it ever was. Even the vocabulary of the unequal contestants has hardly changed.

Attitudes to water exemplify the changing human relationship with nature. Water is such a fundamental for all existence that no animal and no plant escapes the consequences of there being too little or too much.

The accuracy of water as a barometer of human genius or mischief is somewhat reduced by the huge droughts and floods that are precipitated by entirely natural events. On the prehistoric and global scale, long-term changes in climate have moistened or dried out large continental masses, further reducing the scale and significance of anything humans may or may not have done. Nonetheless, the behaviour of water as eroder, spring-feeder or flood-bringer was long ago seen as being linked with human affairs and human actions.

The early Mesopotamian city of Uruq was built on the banks of the Euphrates between present-day Basrah and Baghdad. Because there was no stone or timber in the region, the early Sumerians had to use the river as a canal and fetch these materials from further north, beyond the flood plain. The legendary king of Sumer was Gilgamesh, who is thought to have led one of the first of such expeditions to the headwaters of the great rivers about 5,000 years ago. The cedar forests that once surrounded the sources of the Tigris and Euphrates were seen as under the protection of a deity called Enlil, who symbolised the forces of nature and mastery over the fates of humans. Significantly, his principal weapon of retaliation was floods: as early as 3000 BC a connection was made between attacking forests and being visited by floods. Legends in which attacks on the wilderness were seen as dangerous and an invitation to retribution are extremely ancient and could conceivably have represented vestiges of pre-agricultural beliefs.

By 400 BC legend and symbol were no longer the only language in which man's place in nature could be explored. Plato (427–347 BC) examined the landscape of Attica with an analytical eye and cited what we would call historical or archaeological evidence to back up his observations. In a dialogue, he put the following words in the mouth of Critias:

> What now remains of the once rich land is like a skeleton of a sick man, all the fat and soft earth having wasted away, only the bare framework is left. Formerly, many of the present mountains were arable hills, the present marshes were plains full of rich soil; hills were once covered with forests, and produced boundless pasturage that now produces only food for bees. More-over, the land was enriched by yearly rains, which were not lost, as now, by flowing from the bare land into the sea; the soil was deep, it received the water, storing it up in the retentive loamy soil; the water that soaked into the hills provided abundant springs and flowing streams in all districts. Some of the now abandoned shrines, at spots where former fountains existed, testify that our description of the land is true.

Contemporary archaeologists confirm that the substance of Plato's description was accurate. One thousand kilometres further east, in what is today Lebanon, there had been a rapid depopulation around 700 BC which was due to a collapse of agriculture. The first stage of the decline was a stripping away of all the trees. Over a period of about a hundred years joists and beams in houses got smaller and smaller until, in the absence of wood, arched stone structures had to be devised. The main mortar and plaster used in building was quicklime, the manufacture of which burnt up still more trees. Goats consumed all regrowth, until there came a point where the soil eroded so fast that the fields just wasted away and people had to abandon the area.

The exit of another population has been documented around Lake

Patzcuaro in the highlands of Mexico. The first evidence for serious erosion here was found in a layer of lake-mud sediment dated to about 1600 BC. By 400 BC a dense population all around the lakeshore and foothills was expanding up into the steep pine-covered hillsides. Vast quantities of red soil (the early layers filled with pine charcoal fragments) washed down the slopes and forced the people to abandon all the lake's northern shores. The lake deposits then reveal a pause followed by an even more intense phase of erosion and deposition in the lake's bed between AD 1000 and 1600. The effects of such degradation can be lasting. The Patzcuaro littoral is still gullied and bare with many mud fans and the lake is becoming eutrophic.

It was not generally appreciated until recently that serious degradation would be detectable at such early times. Furthermore, it is widely assumed that pre-agricultural societies maintained an equilibrium with their environment. Although there may have been ecosystems that could withstand quite intense levels of exploitation, this was probably due to more stable climates combined with in-built flexibility on the part of animals and plants. In any case, the fact that a foraging lifestyle was maintained for over 200,000 years may have given a false impression of ecological stability when viewed from our own distant perspective. It is certainly insufficient support for a model of studied prehistoric virtue. The idea that our hunting ancestors maintained a deliberately managed equilibrium is more a stick to belabour consciences than an expression of knowledge. Although hunters would have known the habits of their prey intimately, there is little to suggest that their planning was anything other than observant opportunism and their restraint a plain prudence that normally avoided obvious waste and deliberate overkill; nor were they possessed of exceptional foresight. In general, they lived within ecosystems where regulation of numbers was more a property evolved by the prey than a product of prescient self-control exercised by the predator.

As for regulation of human numbers, that would be more likely to have been a long-term oscillation at a regional level. Where human populations exceeded the carrying capacity of the land, birth-spacing, starvation and fighting would have been the normal stabilising mechanisms. Equilibrium would have been more like a pendulum than a balance, but the amplitude of its swings would have varied very greatly.

Where pressure on the women forced them to space out their children very widely and few infants survived, oscillations in population might have been extremely subtle and spread over many generations. In other areas, boom and bust in prey populations could have induced similar extremes in their human predators. Migration or enlarging the range of foods were essentially short-term alleviations, possible only where numbers had yet to catch up with space and resources.

If prehistoric *Homo* succeeded in living within his means it had less to do with his sapience than with limitations imposed upon him by the quality of his technology and by the recuperative powers of his prey. The interface between hunter and hunted, consumer and consumed, was always the tools and techniques by which food was got. The additive effect of these techniques was a larger impact on the ecosystem which would ultimately affect the entire landscape.

In this respect, the first tool to have a serious impact was fire. At first very restricted in its use it eventually would have become an important hunting tool. The long-term effects of 'cleaning the land' (a widely used phrase among the inhabitants of fire-climax vegetation) were to alter and generally simplify both the plant and animal community.

Fire-climax vegetation types are so widespread and familiar that the fact that humans have been the main agency responsible for their success and extent is sometimes forgotten. While most of the plants able to withstand fire evolved long before humans learnt to kindle it, their pre-eminence from that time onwards (and long before the development of agriculture) was a by-product of the hunters' subsistence techniques. This alteration of vast tracts of land to suit human hunters clearly put many other animals at a disadvantage. More than half the surviving antelope species in Africa are leaf and herb eaters (but many are rare or localised), some are versatile feeders and it is a minority that are exclusively grazers. The grazers have been greatly favoured by human activity and most of the famous savanna animals belong to this class – hartebeest, topi, wildebeest, kob and reedbuck as well as other grass-eaters such as zebra, warthog, grass rhino and various rodents.

Fossils confirm that grass-eaters greatly increased in importance towards the end of the Pleistocene. Yet in the absence of fire, grass would grow on waterlogged soils, on certain cold uplands, on arid steppes and deserts and as a subsidiary plant beneath the woodland canopy. The inescapable conclusion is that humans have shaped the savanna eco-systems of Africa in a quite fundamental way. The species that were unable to cope with the land being 'cleaned' declined or died out long ago. The largest and least versatile species such as the various giraffe, pig, elephant and hippo species were probably taken out first. The animals that are left are the result of natural selection. They have survived and even flourished, because they can cope with fire and hunting even when this comes on top of changes in climate. Human hunters have learnt how and where to harvest these animals; they may know a great deal about their habits and avoid overkill, but the main credit for the animals' survival lies with the animals' intrinsic vitality, rather than with their hunters' ability or wish to conserve them.

It is generally more plausible that prehistoric hunters pushed their technology and their luck to the utmost. If vulnerable species became extinct in the process, hunters would have turned to other sources. A big empty niche is often filled or partially filled by one or more smaller animals that can withstand pressure that large ones could not. Herbivore sizes have declined (and continue to decline, as the last elephants, rhinos and giraffes are taken out) and even the surviving larger animals tend to decline in average body size over time.

In the earlier phases of development the only serious impact humans could make on the environment was through sustained killing of large, slow-breeding animals to the point of extinction and the modification and perhaps extinction of various plant communities that had lost their dispersers or were sensitive to fire. Direct impact on plants through overcutting or overcollecting would have come much later. Breaking the ground with digging sticks, picks or ploughs was also a much later development but both would have been closely linked with the use of fire. Once again populations of farmers would have tended to push out to the furthest capacity of their technology, to the outermost margins of cultivable land and to the limits of the resources they were exploiting. Hunters have been less severe on landscapes than farmers but both have gradually perfected ways of living off a depleted range of survivors.

The pattern has been taken to its furthest extremes by farmers. As scientific research (and Plato) have shown, hills and mountains have been repeatedly reduced to their geological skeletons, yet people still live there. Fewer and fewer species can cope with depleted, rocky soils, rapid run-off and harsh contrasts between seasons. Nonetheless, such species do exist and many of them yield a living to a variety of subsistence cultures. If credit is due for survivorship, first spare a thought for the plants and animals that have endured the so-called management skills of humans over such long periods of time. Consider how admirable are the many hardy plants that, together with sheep, goats and donkeys, make life go on in the ruins of Lebanon or Arcadia. If Plato could lament an already ravaged Greece in 400 BC it is truly a wonder that Mediterraneans can still find blessings in an inheritance that was squandered so long ago.

This is a consistent pattern that is discernible in many human subsistence economies – hunting, gathering, fishing and farming. Because foragers tend to be fewer in number, more remote in time and technologically primitive, their ecological impact is dwarfed by the massive degradation caused by farmers, and minuscule compared to the army of rip-off merchants that have invaded the forests and fields of today's world. It is important, nonetheless, to recognise that the differences are primarily ones of scale – scale of human numbers, and scale of technology. Why make a virtue of prehistoric man's inability to do worse than he did by the fauna and flora of his times?

Sentimentality should be no part of the effort to define prehistoric relationships with nature. The search for underlying patterns demands that the relationship should not be trivialised. If we are to improve on our self-appointed task of managing a planet, self-criticism should at least acknowledge that our failings might be endemic. It is not just we ourselves but out fathers too that are on trial.

Present efforts to name and record the distribution of species and to understand the ecology of natural communities are worthwhile in their own right. They are also, in a global perspective, the inventory of an inheritance. If 200,000 years of management and mismanagement have shaped that inheritance, both the process and its outcome should be explored.

Of course, it is now well known that ecological events have shaped history. For example, there is increasing evidence to show that many conflicts, from the most localised to the more famous battles and wars of recent history, were ultimately about resources. In the Sudd region of the Upper Nile, tribal wars have nearly always coincided with years in which there was a severe shortage of grazing. People dependent on cattle must have access to pasture and will fight to get it or defend it. If desperate enough they will even fight their own kin. There is a sense in which the wars between European imperial powers are comparable with Dinka and Nuer battles. If imperial grazing grounds or stock routes across the sea were threatened, the European tribes were all too ready to go to war with one another.

Earlier still, droughts or longer-term declines in the agriculture and trade that supported some Mediterranean states are thought to have influenced many of the wars and changed the fortunes of regional powers. Exhausted soils, eroded slopes, and the collapse of their rural support base are thought to have caused even more devastating disintegration in the cities. The fall of Phoenicia, Greece, Rome and many other Middle Eastern and Mediterranean civilisations may have a decisive element of ecological as well as cultural breakdown at their root.

The fact that drought and crop failures can cause tribal skirmishes is well enough recognised. What has been forgotten is that really large-scale shifts in climate or persistent underproduction in regions of today's world could precipitate very much more dangerous and contagious modern wars.

The 500 years in which western Europe has pushed out into all the continents and islands has been a period of continuous physical expansion. First came explorers, then traders and in many cases invaders and colonists, all backed up by navies and ever more efficient transport and communication networks. This expansion in territory and in the numbers of imperial subjects coincided with unprecedented inventiveness in industrial, agricultural, medical and military techniques. The present time belongs to the tail end of an era of *laissez-faire* expansion.

There can be little doubt that this history and the establishment of an expanding worldwide economic system have created some of the assumptions that underlie the belief in endless expansion and economic growth. Now, in the twilight of that era the twenty-first century has to have new leitmotifs.

What distinguishes our own generation from all preceding ones is that we have the intellectual and technological tools to view ourselves and our activities, our past and our future in the context of natural processes. These process are astronomic, physical, climatic, chemical, biological and cultural. The only way we can escape using up our resources is to recognise that they are all finite. Our society must also learn to regulate its own greed, its own numbers and, in doing so, come to understand, to respect and to live within the limits imposed by nature. Our technology must be turned away from extracting, fighting, consuming every resource and be rebuilt to mimic the processes that make this planet a home for humans. For a start, what can we learn from lemmings?

Rodent populations explode into scurrying ravenous hordes after a series of coincidences allow their naturally high fecundity free rein for several generations. A prolonged abundance of food and cover allows a succession of breeding pulses to multiply the rodents' numbers exponentially. Meanwhile the usual checks and balances – predators, diseases and food shortage – are in abeyance.

Then the cycle closes. Loss of cover, more predators and then diseases. So little food that rat fights rat for what is left.

We too have set out on a lemming path. In April 1992 the world's human population was 5,480 million. For many, many generations we have had a third, ever more powerful parent to shield us from the many cruelties of nature. Drawn further and further out of our biological matrix we have become more and more dependent on an all-embracing but loveless technology to see us through.

Under this impassive influence we have become orphans of our own technology. Although it is a neutral and neutered parent, technology has warded off starvation, disease and the rigours of climate. It has been left to our biological parents to reproduce. They have been relieved more and more of the personal costs of having children. Many technological societies have gone so far that parents can diminish, shed or even renounce their responsibilities as parents altogether. This cuckoo syndrome is a luxury that can only be temporary.

If there is any lesson for the twenty-first century in prehistoric societies it will lie in a return of much fuller and more demanding responsibilities to the men and women who choose to have children. The present trend towards degrading the value and shirking the costs of having kids will have to be reversed. It is to be hoped that the twenty-first century will seek the ways and means of giving a greater value to fewer offspring. This cannot be a mere technical fix but will involve a social and spiritual revolution.

The political and social fortunes of any growing human population can be divorced from their local environmental and biological underpinnings only as long as they are being subsidised from some external source or are drawing on some form of nonrenewing capital reserve. When reserve or subsidy is exhausted the repercussions can be swift and felt very widely. This is especially so when the collapse is linked with conflict (as it usually is, illustrated by civil wars in Somalia, Ethiopia, Sudan and numerous other countries).

The first to suffer are the very young and the very old. TV news broadcasts have made this a familiar scene. The immediacy TV has brought to our experience of others' suffering forces us to choose between switching off and walking away or becoming, in some manner that is entirely new, involved. It is a choice even our children and grandchildren find themselves having to make.

We cannot make a scapegoat of the technological revolution that has pampered us yet passed by the emaciated victims we see on television. It is an extension of what *we* are. If we are greedy and selfish technology will be a faithful mirror. Left to its own dynamics technological and industrial innovation trashes products, places and people. Technology is at once social shredder, racial churn and political furnace. It is for the children of technology to humanise their parent or, like Saturn, it will consume them. Self-made Man and his society will be undone. If the twenty-first century sets out to build a new sense of family it has powerful tools to help in the task. If it doesn't, its antithesis – increasing conflicts between haves and have-nots – is inevitable.

The study of natural processes, so long confined to the laboratory, has now moved on to the broad stage of international politics and raises issues that must engage us in new struggles.

The most fundamental process by which solar energy is converted into biomass is photosynthesis. Another vital plant process is nitrogen fixation. Scientists have examined plants in the hope of breeding super-plants that will increase yields. The immediate rewards for those who breed, patent, distribute or grow improved cultivars can be very great but, like any other increase that enlarges population numbers, it only defers the day when real limits are met. This is because photosynthesis and fixation are also finite. There are natural limits in plants which determine how much energy the process can generate. These limits are intrinsic to what plants are. However, the belief in unlimited economic growth has had some encouragement from plant breeders, who have persuaded their sponsors to believe they can always stay one step ahead of the Malthusian nightmare of a world with too many mouths to feed.

The 'Green Revolution' of the 1960s held out the hope that plant breeding would solve all the world's food problems. Super-rice, super-maize and other improved breeds have greatly enlarged yields but have not proved so useful for the more remote small farmers who cannot afford the essential fertilisers and whose fields are exposed to a multitude of diseases and predators. Genetic engineering and biotechnology have set out to overcome some of these difficulties by implanting protective and productive genes, but here too there are limitations in the techniques (a plant will tolerate only one or two new genes, if any) and a plant can be modified only within parameters that have been determined by evolution, not by human needs and human manipulation.

Japan has a long and continuous experience of selective breeding. Apart from the well-known golden carp and long-tailed cockerels, Japanese breeders have been improving rice strains and rice yields over centuries. Since the Second World War there has been a strong drive to be self-sufficient in rice. Japanese farmers are paid four times the world's price; they have access to specialist services and research and to funds that reinforce the political incentive to maximise production. Despite so many advantages, these, probably the best-served and best-educated farmers in the world, have managed to increase yields by only 0.9 per cent per year, which suggests that they are close to the limits of what rice is capable of. It can be manipulated so far and no further. The doctrine of infinite economic growth eventually meets biological limits here as in every other sphere.

Only if we begin to understand the processes will there be any chance that we graduate from being maladroit apprentices to being competent sorcerers. The metaphor of the sorcerer's apprentice is not entirely appropriate, because it implies the existence of a sorcerer. The sorcerer is supposed to know what he or she is doing while the apprentice is just meddling with forces he cannot contain and understands only incompletely. By this criterion there are as yet no sorcerers, only rampaging apprentices but there are many among the latter who would have us believe they have graduated.

Perhaps this is where there is something of a watershed between the past and the present. Before Hiroshima no excuses and no explanations were made for pillage, conquest and the rape of the earth. Those who had the means to do what they could, and would, went ahead and did it – the consequences were scarcely ever debated and if they were it made no difference. Mass media and the bombing of Hiroshima and Nagasaki have changed that. You do not have to go to Nagasaki and see the broad basin of its busy harbour and bay to compare it with the numbing series of grey photos that line the walls of the memorial museum. At the very spot where the bomb fell there is a comprehensive record of the smoking cauldron it was in 1945. You can see the bomb's instantaneous retribution printed into a small door, like an upright coffin lid; before he could even crumple and crumble before the blast a sentry left his silhouette charred into the door's wood. They have kept shelf-loads of wilted clocks, all stopped at that fateful second, and the melted bottles. You do not have to go but it helps awake one certainty. Like millions of other people you now know that the world is in the hands of more or less irresponsible apprentices and all of us must go on struggling with the implications.

Anxiety about the environment, nature and our children's future has given birth to several large international bodies and countless national, local and special-interest groups devoted to what is now called conservation. Conservation means different things to different people; objectively the word means 'preservation from destructive influences, decay or waste' and 'official charge and care of rivers, forests, etc.' ('Conservancy' in this last sense dates back at least to 1490.)

While it is 'destructive influences' that exercise many supporters of conservation, there is a much larger fear that destructive influences are merely symptoms of an unpredictable genie that has been released from his bottle and loosed upon the world. This perspective has given rise to what is commonly called the environmental movement. Although they share many concerns with conservationists, supporters of this movement generally focus upon the larger life processes and it is in their focus upon *process* that their authority dwells.

The environmental lobby has become one of the most potent political and social forces of our time. It represents the first organised recognition that there must be some principled guidelines for high-tech societies in an organic and vulnerable biosphere. A pioneering and influential statement of environmental concern was a 1987 report, commonly called the Brundtland Report (actually titled *Our Common Future*), produced by the United Nations World Commission on Environment and Development. The nub of this report was that economic and political policies must change to make the environmental dimension central to all planning. Businesses must be sited, managed and have their raw materials and products influenced by environmental considerations. Transport policies must encourage the search for quieter, cleaner and more energy-efficient engines, and slow the demand for private cars. Energy policy must recognise the global climatic threat from a build-up of carbon dioxide (contributing to the greenhouse effect) and bring in cleaner, more efficient sources of energy. Agriculture must become less dependent on chemicals and poisons.

Four years later the Brundtland Report was succeeded by a review of progress entitled *The State of the Environment*, which was commissioned by the Organisation for Economic Cooperation and Development (OECD) for the environment ministers of its twenty-four member states – a group of countries with 715 million inhabitants. This report reiterated the central tenet of its predecessor, that all economic growth should be judged by how capable it is of being sustained. *The State of the Environment* also identified the tradition of selfish short-term opportunism as a central problem. At every level of policy economic decisions must be made in a way 'that provides for the needs of the present without compromising the ability of future generations to meet *their* needs'. This criterion automatically brings into question the western credo of endless economic growth. The link between economic growth in the big and successful Western nations and degradation in both their own and other countries was made very clear:

> The inhabitants of the O.E.C.D. countries will continue to place a major strain on the world's resources and on the state of the environment through increased consumption and their use and disposal of final products. Consequently, a critical issue is how to prevent the general increase in incomes from being transformed into environmentally harmful consumption patterns.

The report pointed to some successes in less polluted water supplies, reduced sulphur dioxide levels, more recycling and an increase in national parks and reserves but admitted these gains were overshadowed by increasing waste, dwindling sea-fish populations, more nitrogen oxide and agro-chemical pollution. Some of the most worrying examples came from the wealthiest countries. Around Berlin people are having their drinking water delivered by tanker, because the normal sources are so polluted. Excessive use of chemicals and fertilisers are mainly to blame but heavy metals are also polluting wells, springs and rivers. Wells have twenty-five times the EC limits for nitrates, the Havel River is about six times as dirty as the notorious Rhine in phosphates and ammonium and in places it runs with benzol, zinc and the effluent of pharmaceutical factories. The State of the Environment presented detailed examples of damage to the seas, land, water, soil, air and to the animal and plant communities that depend upon these basic elements.

In 1988, an intergovernmental panel on climatic change issued the first official warning about global warming due to the release into the atmosphere of too much carbon dioxide, the 'greenhouse' gas. Changes in temperature and rainfall patterns will alter and shift entire ecosystems over distances of hundreds of kilometres. One effect will be rising sea levels due to melting ice. In general, there will be a massive dislocation of the world as we know it. It has been estimated that a 5°C rise sustained over a very long period was all it took to melt the Ice Cap and take the Earth out of the last Ice Age. Instruments in space are currently monitoring the Earth's radiation and charting this acceleration.

It is not necessary to be apocalyptic. The tribal wars in Sudan that were mentioned earlier may have been localised but they were serious enough for the contestants. In good years the warriors (often closely related to one another) had been the best of friends, yet the exigencies of survival forced them to fight one another. Large-scale shifts in climate and vegetation will have the same effect in a world that has lost the flexibility of nomadism. As each nation's resources get depleted they are less and less likely to be generous to their neighbours, and it is in this context that fierce regional wars could result when crops fail. Even if climatic change is not due to human agency it is such a natural feature of our prehistoric past that international programmes to study the implications of climatic change should have been set up years ago. Such is the short-sightedness of our species.

Our future survival depends upon fundamental changes in industry, society and behaviour, and demands for change will become stronger and stronger as the gaps between rich and poor widen. What form the demands may take is less predictable than the certainty that present patterns of agriculture must change drastically. Land must be used less wastefully and dirtily and, whenever possible, worn-out land must be rehabilitated instead of continuously expanding the area under cultivation. Large areas producing small quantities of expensive or luxury products or those using harmful methods will have to give way to cleaner, more economic and intensive methods. A more disciplined and ecologically sensitive approach to land-use will have to be developed, which takes more account of the natural range of soils, water regimes and climate of a region. In most countries the idea of land-use zoning is already accepted, but where and how much land is given over to what activities will always depend on local social traditions and political attitudes.

Already a few progressive countries have accepted the idea that every vegetation type or ecological community that is natural to a region should be represented in the pattern of land allocation. These countries are seeking to establish nature reserves (of various types and sizes) as an integral part of every landscape, no matter how intensively the rest of it is being used. Less sensitive cultures may take more persuading to this point of view, but a better-educated youth may well change that. For more and more young people, conservation of nature is a litmus test of the quality of their society.

Many new organisations with evocative names have sprung up, such as Friends of the Earth, Greenpeace, Survival International, Living Earth, International Water Tribunal. Greenpeace has had some spectacular campaigns, notably a persistent and successful assault on the whaling industry.

In 1961, when whaling was still at its height, the International Whaling Commission, at that time a pseudo-scientific Front organisation for whalers, issued permits for the hunting of tens of thousands of whales. In that year the catch records showed that most of the largest prey species, the blue whale, were immature. Three years later, the vast Japanese whaling fleet, with 100 large ships scouring the entire southern ocean, failed to find a single blue whale.

Here, in the Antarctic Ocean, the whalers were replaying the old prehistoric drama of hunters hunting their prey to extinction. As in all previous extinctions, the hunters were using a technology that the animals had no defence against. Like their predecessors, the hunters reacted with 'too bad, what's the next size down?' And they turned to larger numbers of smaller whales. However, the threat of extinction of the world's largest mammal received massive publicity when Greenpeace launched its first anti-whaling mission in 1975. Whale stocks are now showing the first signs of recovery and it might seem difficult to see whaling ever again starting up in a big way. However the resumption of whaling will only be deterred by massive and sustained public pressure.

Now fish stocks are dwindling alarmingly. Fishing fleets are vacuuming immense areas and as the catches diminish with one technique new methods are brought in with even more deadly effect. Among the most destructive and wasteful of all are the drift nets, which are tens of metres deep and many kilometres long. Drift nets catch all the larger sea life indiscriminately – whales, seals, dolphins and fish. Greenpeace is rightly campaigning, and with some success, for the total outlawing of such destructive techniques, but the usual short-term arguments are brought in – there will be massive loss of employment, firms will go bankrupt and, above all, governments will lose votes. What the environmental movement seeks to turn around is this last factor. If enough people can be educated to see where such primitive consumption of resources is taking us, they may vote greedy and unprincipled governments out of power. Unfortunately, democratic accountability and environmental steadfastness are still exceptional and some of the most influential governments and publics remain in the dark as far as their responsibilities to the environment go.

Although these organisations depend upon spectacular campaigns and emotive advertisements, the best of them appreciate very well that it is identifying the *nature* of what is wrong that earns them enduring respect. The most successful environmental groups now employ scientists who set out to show where natural processes are being undermined. They identify blackspots, they assemble a dossier and then advertise to the world at large what has gone wrong, giving a bare minimum of suggestions for what needs doing to put it right.

Had governments, companies and societies at large the authority they assert and the wisdom and fairness they claim, there might never have

323

been the need for confrontation. Chemical industries would have avoided poisoning their workers and clients, governments would have found a fair balance between field and factory, town and woodland, work and play. But they have not. The environmental groups are the logical reaction to what the bluntest of them call 'one hell of a cock-up by governments and industry'. I contend that the rise of the environmental movement marks a major turning point in human history.

As our populations have expanded in every continent and eventually into marine habitats, we have consumed and degraded natural resources wherever we have gone. Self-interested opportunism has been a fundamental trait all along. It may have been a motor for action and exploration but it is an inadequate axiom on which to base a political credo, such as permanent growth, nor is it a recipe for survival. If we have been able to work out how both physical and living systems 'work', there is now a new subject that the environmentalists have set up for serious study. How do we live in the world without consuming it? It is a practical question demanding both intellect and techniques to solve, but there is no escaping the fact that it brings environmentalists into direct confrontation with just about every major establishment in the world: governments, banks, multinationals and other big business. These bodies continue the prehistoric tradition of asset-stripping. Examples of the way in which these institutions undermine the environment can be found in the workings of GATT (the General Agreement on Tariffs and Trade).

The 1990 meeting of GATT established rules to extend free trade among 105 countries. One potential of GATT rules is to license plunder, another is to encourage pollution, and to label sustainable development and conservation as protectionism. GATT's sweeping powers can force small countries to open up their precious resources to international trade. By refusing to discriminate between products that are made with less destructive and less polluting methods and those that are not, GATT offers decisive advantages to those industries and nations that can bypass the real environmental costs of their products.

As an enlarging human population occupies more and more of the habitable land, opportunities for large-scale adventurism on dry land have diminished. At the very least they involve the complication of opinionated publics and landowners. There remain two short pathways to massive profits and both involve enormous biological costs in the disruption or destruction of natural ecosystems. Timber from tropical forests and meat from open (or opened-up) rangelands offer fast and rich returns for minimal investment and with very few labour or social complications. For banks, politicians and businesses in the fast lane, this makes these two forms of asset-stripping very attractive. Felling of timber and conversion of wild lands to beef-lots can take place in remote regions, where outsiders can be kept off and scrutiny avoided for the short time it takes to 'convert' or to fell.

Africa's savannas being the primary habitat of mankind, their fate is of obvious concern. Like tropical forest, the savanna is a target for some of the most destructive so-called 'development' that is taking place any-where on earth. The cowboys and pirate loggers are not picturesque outlaws in remote ranches and log cabins; they are the Great and the Good, ministers, bank presidents and captains of industry.

Free trade in beef, sugar and other commodities encourages further clearing of forests and wild lands, as impoverished tropical countries, in debt to bodies like the World Bank and to the industrialised nations, capitulate to demands for new markets. Even the economists and politicians of such client countries have tended to accept the rich countries' credo of endless economic growth and sometimes work against their own long-term interests. For example, it requires very considerable psychological readjustments and some education in ecol-ogy to see wild animals and their habitats as valuable resources rather than obstacles to quick riches. Such changes in perception have only just begun to develop in a few countries and older habits are still prevalent. Primitive attitudes to the wilderness have been encouraged by banks and businesses, because what else is the tradition of making a fast buck if not primitive?

It is popularly promoted that poachers are the main threat to African wildlife, but poachers are generally marginal to the onslaught being made on wildlife habitats by livestock and timber industries that are being encouraged to expand, and expand again, by GATT, the World Bank and national governments. The extinction of some of Africa's remaining ungulates and of the spectacular carnivores that prey upon them could ultimately be the responsibility of international markets that promoted the conversion of viable and very ancient ecosystems into chemical-dependent beef-lots.

GATT's provisions allow grotesque and aggressive claims to be made. For example, super-powerful Japanese logging companies have bought into American firms and, invoking GATT rules, complain that the conservation of ancient forests in northwestern USA unfairly deprives them of timber! Lesser nations, especially those deeply indebted to their main sources of vehicles and industrial and electronic goods, are bludgeoned into granting grossly inequitable licences to strip huge tracts of country of their forests. Papua New Guinea and Indonesia have been among the many victims of this process. Some time ago, the latter country banned the export of raw logs, because it wished to conserve a valuable and slow-growing product. An incidental benefit was that this encouraged its own carvers and furniture-makers. Free trade, however, requires that foreign and domestic industries be treated equally, so aggressive logging interests are using Article 1 of GATT to demand access to Indonesia's raw materials.

Asset-stripping has always been the short route to riches, especially in the tropics. What is new is that the end of it all is suddenly in sight. Just as the blue whales got rarer and rarer and the cod smaller and smaller, the areas where loggers can take out the big trees are shrinking by the year. As this shrinking proceeds, the superstructures – the mills, the markets and all the middlemen – feel the squeeze. Where the forests are all consumed they simply go bust, but most go on putting more pressure on smaller forests or switch to smaller and less valuable types of timber.

In most countries with forests, loggers and pulpers are a powerful lobby that puts governments under continuous pressure to maintain the loggers' control over the forests and, whenever possible, to allow them to expand into virgin forests where the largest and most valuable trees remain. So long as the appetite of both local and overseas markets could be satisfied, key areas of wildlife habitat and the traditional homes of small forest tribes were accepted as being exempt from logging and, in most cases, legislation protected such areas in the name of conservation and ethics. As the wildcat loggers' era draws to a close, protective legislation, the pro-tribal lobby and conservationists have become the real bugbears for logging interests. These apparently flimsy barriers seek to keep what is still regarded as a respectable 'industry' from the last major sources of trees. Its machinery and markets were built to consume those trees, and ministers who control forests can always be persuaded of the economic benefits of logging. Sometimes logger and minister are the same person.

The island of Borneo has long been home to a variety of tribes that lived by hunting and swidden agriculture. The island was partitioned in colonial times and the northwestern quarter, known as Sarawak, now a state of the Malaysian Federation, was the territory of the Kayan, Kelabit, Penan and Iban tribes. During colonial times trade was largely in the hands of entrepreneurs foreign to the island, and at this time a logging business called Limbang Trading was set up. This remained a modest concern until two events in neighbouring countries changed everything. Indonesia decided it could process its own timber and closed its logging concessions to foreigners. The Marcos regime, having allowed most of the Philippines to be stripped of their forests, fell and a similar ban on the export of timber was brought in there.

A huge appetite for wood, especially in Japan, was suddenly in need of new supplies. The Japanese trading multinational C. Itoh moved in to make a partnership with Limbang Trading. In 1987, it was revealed that Japanese government money earmarked as aid for Malaysia had been spent on building roads to Limbang Trading's logging concessions. Far from denting the reputation of Limbang Trading, this injection of funds merely put more power to the elbow of its owner, James Wong. Wong combined a controlling stake in this firm (with concessions to log 300,000 hectares of Sarawak) with being Sarawak's minister of the environment! To put such a contradiction in its local perspective, the New Straits Times in 1987 pointed out that virtually every representative in the Sarawak state legislature and many of their relatives had become millionaires through timber concessions handed out by the former chief minister Tun Rahman.

The unpoliticised tribes living in the forests have been the victims of this gold rush. The loggers' heavy equipment and roads were driven over the gardens, rice paddies, cemeteries and water sources of the locals. Their rivers are now so full of sediment that fishing has been affected. Wild harvests of animals and plant material are a fraction of those in the past.

This is the background to a trade that would outrage ordinary Japanese consumers were they to know that the 15 million cubic metres of timber they import from Sarawak were obtained in such a way. A 1990 opinion poll revealed that only 8 per cent of Japanese think that environmental protection should be subsidiary to economic growth and still fewer would approve of the impoverishment that their own and other multinationals are inflicting on tropical habitats and people.

Faced with declining yields in almost all their most important commodities – bush meat, fish, nuts, resins, bamboos and rattan – the Penan and other tribal groups took to erecting manned barricades to protect their forest. They have endured hundreds of arrests and acts of intimidation but this has served only to increase their resolve. A Sarawak/Penan Association has been formed to demand that the logging stops. Among their petitions to government is an interesting detail. They ask that there should be a ban on any further exports of their declining stocks of rattan unless it has been woven into products. Under GATT regulations this humble proviso would be illegal. It is in an accumulation of many such impositions that captains of industry undermine the efforts of small countries and communities to achieve independence, self-sufficiency and sustainable development.

The onslaught on rainforests has generated one scandal after another, and the underlying motive for these repetitious exhibitions of venality is obvious: tropical timber represents one of the world's largest and most valuable 'free' commodities. It is only access that can be a problem. In Sarawak the Penan were the first to protest; in other parts of the tropical world, environmentalist and green lobbies have created minor obstacles to access.

Green acquiescence and actual financial support were built into another major rainforest logging scheme, the Tropical Forest Action Plan or TFAP. Impetus for the formation of this plan came from growing public awareness during the 1960s and 1970s that the forests were disappearing. The catalyst for action was a realisation of the *pace* of destruction. Scientists estimated that 11 million hectares were felled in 1980 alone. At this rate, it was only a matter of time before tropical forests together with all their fauna and flora would be gone. The trend had to be reversed.

The TFAP aimed, in its own words, 'to end the present state of alarm about the destruction of rainforests'. Yet in the five years following the plan's inception, between 14 and 20 million hectares of forest were felled each year, a total of 85 million hectares, and many logging enterprises were been directly aided by TFAP support. Far from reversing the trend TFAP could be indicted for accelerating it.

Some TFAP enterprises may have irretrievably impoverished future generations. For example, Cameroon is a country where forests are sustained by up to 10,000 mm of rain per annum. Under this sort of rainfall nutrients can only circulate through living plants and animals – all the rest are swept away by sheets of water. Local people, especially the Baka Pygmies, have devised their own ways of making a living in close relationship with the forest. The TFAP helped set up a massive asset-stripping exercise in which a 600 km arterial road was to be cut, mainly to assist logging. The plan boldly asserted its aim to make Cameroon the most important African producer and exporter of forest-based products from the start of the twenty-first century. But its forest soils will be leached and barren once the cover is gone.

Such dangerous and destructive mega-schemes, promoted and assisted by funds donated by people anxious to *save* rainforests, raised the immediate question, how could it have happened? Part of the answer lay in the people and organisation charged with the plan's execution, but a larger part lay in the devious ways people seek access to resources.

In June 1984, the World Resources Institute, WRI, a Washington-based body that researches environmental policy, responded to public concern by issuing a draft of a report entitled *Tropical Forests. A call for Action*. Naturally such an invitation was directly relevant for the main United Nations body concerned with forests, the Forestry Department of the Food and Agriculture Organisation. FAO technocrats saw this as a threat, as was clear from their refusal to participate in the plan, and in October 1985 they rushed out their own *Tropical Forestry Action Plan*.

By 1987 a makeshift alliance had been put together that attempted to amalgamate the WRI and FAO plans. This version of the TFAP was funded by the United Nations Development Programme, the World Bank, the Rockefeller Foundation and other well-wishers to the tune of $8 billion.

A central premise on which the TFAP was founded, funded and accepted was that logging should be conditional upon sustainability and, even more important, upon broader assessment of the social and ecological effects of the plan's action. That the plan was economic and technical in its scope was evident from the start. This was even expressed in the way it classified forests as 'fallow, productive, unproductive, un-managed' and 'logged-over'. The environmental concerns that gave the TFAP their qualified blessing should have registered that there was no serious attempt here to understand the living dynamics of tropical forests; only a haste to get access to the timber under a masquerade of sustainable development and resource security.

Sustainability requires knowledge of the processes by which energy is generated and cycled through an ecosystem, and management, if possible, can only mimic or assist wholly natural renewal through understanding these processes. The repercussions of change in such a complex ecosystem are known to be very far-reaching both for forest people and those living downstream but the social and ecological studies that had been envisaged as central to the plan (and intended to be designed and executed by independent professionals) were never done. Instead, these components became the trifling condescensions of TFAP functionaries. Such assessments as were made were mere rubber-stamping exercises conducted by short-term hired consultants. Clients merely used the TFAP to fund the expansion of their timber industries, while peripheral surveys of fauna and flora sometimes provided the fig leaves supposed to cover the nakedness of their exploitation.

The most immediate outcome of this scandal was to tarnish still further the image of its sponsors. Three of these, the FAO, UNDP and World Bank, are among the largest financial and development agencies in the world. It also deeply embarrassed the national governments that had thrown their taxpayers' money into the TFAP (Britain alone gave £100 million a year). Intended as a major showpiece of the talents, skills and financial beneficence of the industrial world in tackling global problems, the TFAP turned out to be a shabby display of managerial incompetence, mis-use of resources and secretive wheeling and dealing. The magnitude of its failure was compounded by the grandiose propaganda of its founders.

These huge institutions of the modern world are fundamentally flawed in the way they go about things. As they get bigger and as more and more political power and technological hardware get concentrated within these oligarchic structures, the more their directors and executives threaten to become new sorcerer's apprentices for the twenty-first century.

Who gets to control or use technology will continue to vary enormously from country to country. It can be seized (up to a point) by the Galtieris and Idi Amin Dadas. It can be inherited, it can be bought or educated for and it can be pursued as the reward for individual ambition through any one of modern society's many institutions, from politics and the armed forces to industry and trade.

With high technology falling into the hands of such an ill-assorted gang, setting guidelines and putting limits upon the innumerable and often dangerous side effects of technology is going to assume ever greater importance in the future. Access to technology, techniques and processes needs to be kept open to all, yet responsibly controlled in the interests of the world's life processes of which we are a mere part.

In this respect a new and very important concept is that of an 'Earth Council'. Such a body would promote up-to-date analyses and gather information on every aspect of the environment. It would disseminate such knowledge as well as follow up the practical implications of new knowledge and developments. Strictly independent from commercial and government agencies, it would aim to advise and, if necessary, censure environmental transgressors, no matter how powerful or influential they might be. The Sarawak and TFAP scandals serve to emphasise the need for such global environmental watchdogs. It will be important however that the council has teeth and is not one more international placebo.

An organisation of this nature has never been attempted before (except perhaps in the field of human rights, where Amnesty International might be the closest parallel). What is needed is a global network of just such international institutions that will be independent, principled, respected and strong.

In these organisations loyalties will have to shift away from shallow short-term nationalism towards the interests of a larger human family and its descendants. In this shift scientists, journalists, technicians and bureaucrats must ensure that issues such as the environment, resources, human rights and nuclear power are debated openly, not settled or stifled by entrenched cliques or secretive technocrats.

Human existence in all its variety, ingenuity and creativity is tied in with technology as it always has been. Technology is what defines our time and place today and every day, back into the most distant beginnings of humanity. There may never have been a time when technology, however humble, was not widely used but, as it gets more and more comprehensive in its effects, it is certain that we must make it subservient to the biological health of this planet. We are rooted in the earth and it is to the earth that we shall return no matter how sophisticated the shields and mirrors that blind us to that ultimate destiny.

Glossary and Abbreviations

There are several terms which become cumbersome when used with any frequency and one of these is: 'Years before the Present Time'. The abbreviation 'BP' is the main time unit used in this book. To circumvent many of the clumsy names, such as 'archaic modern human' to describe the stock that was intermediate between Erect and fully Modern humans I have adopted the taxonomy that calls them 'Heidelbergs'; I also prefer 'forager' to the clumsy 'hunter-food-gatherer'; furthermore, generalised gathering and small-scale opportunistic trapping and hunting (as practised by all recent foragers) is decisively different from living off large mammals. The terms 'mega-hunter' and 'mega-hunting' serve to make a distinction.

The tropical shore-gatherers that are such an important part of this story also need a distinct appellation. Although they must have had watercraft they were not strictly fisherfolk (because living off sea fish was a rather late development). To distinguish the beach dwellers from their riverine predecessors and all other neighbouring economies I have borrowed the title of an obscure and now extinct Namibian group, 'strandlopers', and transferred this name to the much larger and more widespread prehistoric economy.

I also had to find a comprehensive term for the early people that colonised Southeast Asia, Melanesia and Australia. Indonesian, Melanesian and Australoid all have restricted and modern meanings that will not do. The Banda Sea lies east of Indonesia and north of Australia. As the centre of their range it serves to name the prehistoric Banda people.

Acheulian Toolmaking tradition 1.5 MY–150,000 BP in Africa and Eurasia typified by hand-axes.

Adaptation Evolution of features that aid organisms to live and reproduce in their environment.

Aminoacids The 20 chemical components of protein, each a unique building block of carbon, hydrogen, oxygen and nitrogen.

Anthropoid A class (suborder) of primates that includes monkeys, apes and humans.

Artefact Any object fashioned by humans.

Aterian Middle Stone Age (Palaeolithic) toolmaking tradition. North-west Africa.

Aurignacian A toolmaking tradition beginning after 40,000 BP in Europe. Typified by worked stone, bone and antlers – long retouched blades and bone points.

Australopithecines A class (sub-family) of hominids or man-apes, restricted to Africa and recorded from over 4 MY to 1.3 MY.

Banda A prehistoric Modern human population centred on the Banda Sea, inhabiting many of the islands and shores of the equatorial Indo-Pacific.

Biota All the animals and plants in an ecosystem or region.

Bipedal Two-legged walking.

BP Before the Present. A crude measure of prehistoric and historic time that uses 1950 as marker date for the 'present'.

Capsian Toolmaking tradition in North Africa.

Caucasian An obsolete term to describe Modern humans from north of African 'negroids' and west of Oriental 'Mongoloids'.

Chromosomes Elongated bodies of DNA within the nucleus of every cell, carrying inherited characteristics of the whole organism.

Cranium Brain case of the skull.

Cro-Magnon Variety of humans occupying Europe between 30,000–12,000 BP.

Core Stone matrix from which flakes are struck off by the stone knapper.

DNA Deoxyribonucleic acid. The double helix molecule that stores and passes on chemically coded genetic information from generation to generation.

Ecology Study of factors that govern how organisms interact with one another and with their environment.

Ecosystem The organisms of an ecological community together with their physical environment.

Erects, *erectus* A species of human that preceded Modern humans.

Ethiopian realm Biogeographic region comprising continental Africa and the Arabian peninsula.

Evolution Process by which animals and plants change in form over many generations in order to adapt to their environment and its biotic community.

Fauna Collective term for all animals in an ecosystem.

Flake A very common stone artefact struck from a core.

Flora Collective term for all plants in an ecosystem.

Florisbad A late Pleistocene fossil site in South Africa.

Fossil Any preserved evidence of prehistoric plants or animals.

Gene A unit of inheritance. A segment of DNA within the chromosome with coded information for the synthesis of particular proteins.

Gene pool Sum total of genes in a population.

Genus Group of closely related species sharing a common ancestor and generic name.

Glaciation Ice Age cold periods in which ice caps and glaciers expanded.

Habilis, Habilines A group of fossil hominids and the earliest *Homo*. The population named 'Handymen'.

Habitat Environment that supports life.

Hadar Major fossil site in the Ethiopian Rift Valley.

Hand axe Common stone tool with two working surfaces, normally 8–30 cm long and typical of the Acheulian period.

Herbivore Plant-eating animal.

Heredity Transmission of characteristics from parents to their descendants.

Holocene The last 10,000 years.

Hominids Zoological family. Any member of human lineage since its separation from the apes.

Hominoid (Hominoidea) Zoological super-family which contains all apes and humans, living and fossil.

Homo Genus in which all living and fossil humans are grouped.

The Horn, (Horn of Africa) Northeastern tip of Africa, i.e. Somaliland.

Human Generic term for all members of genus *Homo* and characteristics pertaining to them.

Interglacial Warmer period between cooler glacial periods.

Interstadial Short relatively warmer interlude within a longer glacial period.

Isimila Early Stone Age site in southwest Tanzania.

Java man Erect human from the island of Java first called *Pithecanthropus erectus*.

Jebel Irhoud Fossil site in northwest Africa with landmark fossils of archaic Modern humans.

Kalahari A sub-desert region in southwest Africa.

Kalambo (Falls) A rich fossil site at the south end of Lake Tanganyika.

Knapper, knapping The maker, making of stone tools.

Khoisan Original inhabitants of southern Africa. Formerly called Bushmen, comprising Khoi in the Cape and San further north.

Klaasies River Important fossil site in South Africa.

Kung (Kung San) A tribal division of San or Khoisan people.

Levallois A refined technique of stone working by detaching a pre-shaped section of core.

Levant Region on eastern shores of the Mediterranean.

Lucy, Lucies Popular name for an individual of *Australopithecus afarensis*. Extended to distinguish *A. afarensis* from *A. africanus*.

Lupemban A refined stone-tool tradition from equatorial Africa.

Magdalenian A late Stone Age (Upper Palaeolithic) tradition in Western Europe (18,000–12,000 BP).

Maghreb North African Mediterranean littoral.

Malawi A lake and country (region) in southeast Africa.

Mauretania A country (region) in northwest Africa.

Mesolithic Eurasian and north African Stone Age period between the Palaeolithic and Neolithic.

Mechta Afalou A fossil site in north Africa.

Microlith Stone artefacts, usually less than 3 cm long, incorporated in composite tools, i.e. sickles, saws, arrows or harpoons.

Middle Stone Age A stone working tradition in Africa and Eurasia that occurred in a time span from 180,000 to 10,000 BP.

Miocene Geological period between 23–5 million years.

Mitochondrion Self-reproducing structures within the cell which have 37 genes and provide the cell with energy.

Mt DNA Mitochondrial – dNA Rings of DNA which are independent of the nucleus. Only inherited through the egg, not the sperm.

Mongoloid Term commonly used to describe Eastern Asians and Amerindians.

Mousterian Stone-tool tradition in Western Eurasia between about 100,000 and 34,000 BP.

MY Million Years.

Natural selection Evolutionary process whereby best adapted generations tend to survive and propagate their lineage.

Neanderthal A Western Eurasian population of prehistoric humans 150,000–32,000 BP.

Negroid Obsolete term used to describe the peoples of Africa.

Neolithic The New Stone Age. A late tradition in which stone tools, pottery and crop-raising occur together.

Nutcrackers Popular name for the heavily built, large-jawed Australopithecines. Also called *Paranthropus* or Paranthropines.

Oldowan Oldest recognised tradition of stone working and associated with *Homo habilis*, mainly in Eastern African sites.

Olduvai Gorge Major prehistoric site with long series of fossils and human artefacts.

Organism Any living thing.

Olorgesailie A fossil and artefact site in southern Kenya (Rift Valley).

Omo River A valley draining southern Ethiopia and emptying in Lake Turkana with numerous fossil sites.

Palaeobiology Study of past biological processes through fossils, geology and extrapolation from principles and processes.

Palaeolithic The Old Stone Age ranging from approximately 2.6 MY to 10,000 BP.

Palaeontology Study of past life, principally by means of fossils.

Palestine Eastern Mediterranean region.

Paranthropus Generic name (meaning equal to man) for the nutcrackers or large jawed Australopithecines.

Peking Man (Pekin, Beijing) A regional population of erect humans, originally described as *Sinanthropus pekinenssis*.

Photosynthesis Process by which plants convert carbon dioxide and water into energy-rich sugar by means of light.

Phylogeny Evolutionary lineages (often depicted as branches of evolutionary tree).

Pleistocene Geological period 1.6 MY–10,000 MY.

Pliocene Geological period 5 MY–1.6 MY.

Population Group of individuals of same species in discrete area at a given time. In genetics an interbreeding group or pool belonging to a single species.

Primates Order of mammals including humans, apes, simian monkeys and prosimians.

Proteins Organic molecules containing chains of amino acids in specific sequences. Structural material for building many parts of bodies.

Quaternary Term of convenience for Pleistocene and after, up to the present.

Radiocarbon A dating method suited to material up to 40,000 years old.

Relict, relictual Localised survival of group or species after extinction in other parts of its former range.

San The more northerly representatives of a modern South African population group, the Khoisan.

Sangoan An African stone-tool industry making coarse, heavy picks and hand-axes of uncertain dating. Sometimes appears to be contemporaneous with very different, more refined techniques.

Savanna Grass or scrubland plain with scattered trees, wooded grassland.

Scraper Stone tool with retouching on one or more edges (not necessarily used for scraping).

Sexual dimorphism Differences in size or form between males and females of the same species.

Speciation Process whereby new species emerge from a different earlier stock.

Species Group of individuals that can interbreed and are reproductively distinct from other such groups.

Stillbay Middle Stone Age industries in eastern and southern Africa, typically used small lance-like points.

Strandloper Term used in this book to describe a prehistoric culture in which the people exploited coastal and estuarine habitats on tropical islands and seashores (appropriated from a recent population living on the Namib coast).

Stratigraphy Study of the formation, constituents and sequence of geological layers in time and space.

Swartkrans An important fossil hominid site in South Africa.

Tanganyika The mainland portion of Tanzania. Also a deep Rift Valley lake.

Tanzania The East African Republic uniting Tanganyika and Zanzibar.

Taxon (pl. taxa) Group of organisms judged to be sufficiently distinct to be accorded a formal name, i.e. order, family, subspecies.

Taxonomy Study of classification of organisms.

Terra Amata Fossil site in Central Spain.

Torralba Fossil site in Central Spain.

Touareg (Tuareg) A distinct nomadic people of the Sahara.

Tradition Archaeological term for groupings of prehistoric industries or cultures.

Tshitolian A late equatorial African stone-tool industry.

Vedda A distinct people inhabiting southern India and Sri Lanka.

Vertebrate Animals with skulls and backbones.

Yam Vines with large edible roots. Very abundant in many tropical areas.

Zambezi The major river draining southeast Africa.

Bibliography

Aikens C. M. & Higuchi T. 1982. *Prehistory of Japan*. Academic Press, N.Y.

Airvaux J. & Pradel L. 1984. Gravure d'une tête humaine de face. La Marché Sté. *Préhist. Franç.* Tome 81, –7.

Aitken M. J. 1974. *Physics and Archaeology*. Clarendon Press, Oxford.

Akazawa T. & Aikens C. M. (Eds), 1986. *Prehistoric Hunter-Gatherers in Japan*. University of Tokyo Press, Tokyo.

Allen J., Golson J. & Jones R. (Eds), 1977. *Sundha and Sahul*. Academic Press, London.

Allen J., Gosden C. & White J. P. 1989. Human Pleistocene Adaptations in the Tropical Island Pacific. Recent Evidence from New Ireland. *Antiquity*.

Allison A. C. 1954. Protection afforded by Sickle-cell Trait against Subtertian Malarial Infection. *British Med. Journ.* I, 290–294.

Anderson A. 1989. Mechanics of Overkill in the Extinction of New Zealand Moas. *Journ. Archaeol. Sci.* 16, 137–151.

Anderson J. G. 1943. Researches into the Prehistory of the Chinese. *Bull. Mus. Far East Antiquities. Samlingarna* 15.

Andrews P. 1982. Hominoid Evolution. *Nature* 295, 185–186.

Andrews P. & Franzen J. L. (Eds), 1984. The early Evolution of Man with special emphasis on Southeast Asia and Africa. *Cour. Forsch. Inst. Senkenberg*, Frankfurt.

Arambourg C., Boule M., Vallois H. & Verneau R. 1934. Grottes de Beni Segoual. *Archives Inst. Pal. Hum. Mémoire* 13, Paris.

Ardrey R. 1961. *African Genesis*. Collins, London.

Bahn P. & Vertut J. 1988. *Images of the Ice Age*. Windward, London.

Bailey G. (Ed), 1983. *Hunter Gatherer Economy in Prehistory*. Cambridge University Press, Cambridge.

Baker P. & Weiner (Eds), 1966. *The Biology of Human Adaptability*. Oxford University Press, Oxford.

Bar-Joseph O. 1980. The Prehistory of the Levant. *Ann. Rev. Anthrop.* 9, 101–133.

Barnicot N. A. 1957. Human pigmentation. *Man* 144, 114–120.

Barzun J. 1965. *Race: a Study in Superstition*. Harper and Row, N.Y.

Behrensmeyer A. K. & Hill A. P. (Eds), 1980. *Fossils in the Making*. University of Chicago Press, Chicago.

Bellwood P. 1979. *Man's Conquest of the Pacific.* Oxford University Press, Oxford.

———— 1985. *The Prehistory of the Indo-Malaysian Archipelago.* Academic Press, Sydney.

———— 1987. *The Polynesians.* Thames and Hudson, London.

———— 1989. *Cultural and Biological Differentiation in Peninsular Malaysia: the last 10.000 years.* 2nd Intern. Conf. on Malay Civilisation, Kuala Lumpur.

Berndt R. M. & Berndt C. H. (Eds), 1965. *Aboriginal Man in Australia.* Angus and Robertson, Sydney.

Bicchieri M. G. (Ed), 1972. *Hunters and Gatherers Today.* Holt Reinhart & Winston, N.Y.

Binford L. R. 1981. *Bones, ancient Man and modern Myths.* Academic Press, N.Y.

———— 1983. *In Pursuit of the Past.* Thames and Hudson, London.

———— 1989. *Debating Archaeology.* Academic Press, N.Y.

Birdsell J. B. 1967. Trihybrid Origin of the Australian Aborigines. Archaeology and Physical Anthropology. *Oceania* Vol. 2, No 2, 100–155.

———— 1972. *Human Evolution. An Introduction to the new Physical Anthropology.* Rand McNally, Chicago.

Bishop W. W. (Ed), 1978. *Geological background to Fossil Man.* Academic Press, Edinburgh.

Bishop W. W. & Clark J. D. (Eds), 1967. *Background to Evolution in Africa.* University of Chicago Press, Chicago.

Bodmer W. F. & Cavalli-Sforza L. L. 1976. *Genetics, Evolution and Man.* Freeman, San Francisco.

Bordes F. 1972. *The Origin of* Homo Sapiens. UNESCO, Paris.

Bordes F. 1968. *The Old Stone Age.* McGraw Hill, N.Y.

Boughey A. S. 1975. *Man and the Environment.* Macmillan, N.Y.

Bourlière F. 1985. Primate Communication: their Structure and Role in Tropical Ecosystems. *Intern. Journ. Primat.* 6, 1–26.

Bowler P. J. 1984. *Evolution. The History of an Idea.* University of California Press, Berkeley.

Bowles G. T. 1977. *The People of Asia.* Weidenfeld, London.

Boyd W. C. 1950. *Genetics and the Races of Man.* Little, Brown, Boston.

Branda R. F. & Eaton J. W. 1978. Skin-colour and Nutrient Photolysis: an evolutionary Hypothesis. *Science* 201, 625–626.

Brauer G. 1984. The Afro-European *sapiens* Hypothesis and Hominid Evolution in Asia during the late Middle and Upper Pleistocene. *Cour. Forsch. Inst. Senkenberg* 69, 145–165.

Broom R. 1950. *Finding the Missing Link.* Watts & Co, London.

Brothwell D. R. 1963. Evidence of early Population Change in Central and Southern Africa: Doubts and Problems. *Man* 132, 101–104.

Brown J. A. & Price T. D. (Eds), 1985. *Prehistoric Hunter-Gatherers; the emergence of Cultural Complexity.* Academic Press, N.Y.

Brown M. 1976. Some Historical Links between Tanzania and Madagascar. *Tanzania Notes and Records* No. 79.

Bruhes A. M. 1977. *People and Races.* Macmillan, London.

Bulmer S. & Bulmer R. 1964. The prehistory of the New Guinea Highlands. *Amer. Anthrop.* 66, 4, part 2, 39–76.

Bunn H. T. 1982. *Meat-eating and Human Evolution.* PhD Diss., University of California.

Bunn H. T. & Kroll E. M. 1986. Systematic Butchery by Plio-Pleistocene Hominids at Olduvai Gorge. *Current Anthrop.* 27, 331–442.

Burkill I. H. 1953. Habits of Man and the Origins of Cultivated Plants of the Old World. *Proc. Linn. Soc.* 164 (1), 12–42.

Butzer K. W. & Isaac G. L. (Eds), 1975. *After the Australopithecines.* Mouton, The Hague.

Butzer K. W. 1982. *Archaeology as Human Ecology.* Cambridge University Press, Cambridge.

Byard P. J. 1981. Quantitative Genetics of Human Skin Colour. *Yearbook Phys. Anthrop.* 24, 123–137.

Campbell B. G. 1974. *Human Evolution.* Aldine, Chicago.

Cann R. L., Stoneking M. & Wilson A. C. 1987. Mitochondrial DNA and Human Evolution. *Nature* Vol. 325, 31–36.

Carey J. W. & Steegman A. T. 1981. Human nasal protrusion, latitude and climate. *Amer. Journ. Phys. Anthrop.* 56, 313–319.

Caughley G. 1988. The Colonisation of New Zealand by Polynesians. *J. Roy, Soc. New Zealand* 18, 245–270.

Cavalli-Sforza L. L. 1968. Studies on African Pygmies. *Amer. Journ. Hum. Genetics* 21, 252–274.

Cavalli-Sforza L. L. & Bodmer W. F. 1971. *The Genetics of Human Populations.* Freeman, San Francisco.

Cavalli-Sforza L. L. & Feldman M. W. 1981. *Cultural Transmission and Evolution: a Quantitative Approach.* Princeton University Press, Princeton.

Cavalli-Sforza L. L. & Piazza A. 1975. Analysis of Evolution: evolutionary independence and freeness, *Theor. Pop. Biol.* 8, 127–165.

Cavalli-Sforza L. L., Piazza A., Menozzi P. & Mountain J. 1988. Reconstruction of Human Evolution bringing together genetic, archaeological and linguistic data. *Proc. Nat. Acad. Sci.* U.S.A. 85, 6002–6006.

Cesare E. 1968. The Pleistocene Epoch and the Evolution of Man. *Current*

Anthrop. 9, 27–47.

Chang K. C. 1986. *The Archaeology of Ancient China.* Yale University Press, New Haven.

Chang K. C. 1970. The beginnings of Agriculture in the Far East. *Antiquity* XLIV.

Chang S. *et al.,* 1980. *Atlas of Primitive Man in China.* Science Press, Beijing.

Chappell J. 1974. Geology of coral terraces on Huon peninsula New Guinea, a study of quarternary tectonic movements and sea-level changes. *Bull. Geol. Soc. Amer.* 85, 553–570.

Chappell J. & Shackleton N. J. 1986. Oxygen isotopes and sea level. *Nature* 324. 137–140.

Cheyney D. & Seyfarth R. 1990. *How monkeys see the world.* University of Chicago Press, Chicago.

Chomsky N. 1968. *Language and Mind.* Harcourt Brace Jovanovich, N.Y.

Clarke C. A. 1964. Blood groups and disease. *Discovery* Jan., 40–44.

Clark G. 1961. *World Prehistory.* Cambridge University Press, Cambridge.

Clark J. D. 1970. *The Prehistory of Africa.* Thames & Hudson, London.

Clark J. D. (Ed). 1982. *The Cambridge History of Africa.* Vol I. Cambridge University Press, Cambridge.

Clutton-Brock J. & Grigson C. (Eds), 1983. *Animals and Archaeology: Hunters and their Prey.* BAR Series 163, Oxford.

Cole S. 1954. *The Prehistory of East Africa.* Pelican, London.

———— 1965. *Races of Man.* British Museum, London.

———— 1975. *Leakey's Luck.* Collins, London.

Coon C. 1963. *The Origin of Races.* Knopf, N.Y.

———— 1965. *The Living Races of Man.* Knopf, N.Y.

Coon C., Garn S. N. & Birdsell J. B. 1950. *Races.* Thomas Springfield.

Coon C. S. 1982. *Racial Adaptations. A Study of the origins, Nature and Significance of racial variation in Humans.* Nelson Hall, Chicago.

Coppens Y. 1983. Les plus anciens fossiles d'Hominides. *Pontif. Acad. Scient.* 50, 1–9.

Coppens Y., Howell F. C., Isaac G. L. & Leakey R. E. F. (Eds), 1976. *Earliest Man and Environments in the Lake Rudolph Basin.* University of Chicago Press, Chicago.

Crawford M. H. (Ed), 1984. Black Caribs. A Case study in Biocultural Adaptation. *Current Dvts. in Anthrop. Genetics* Vol 3, Plenum, N.Y.

Coursey D. J. 1967. *Yams.* Longmans Green & Co., London.

Dahlberg F. (Ed), 1981. *Women the Gatherers.* Yale University Press, New Haven.

Dale T. & Carter V. G. 1955. *Topsoil and Civilisation.* University of Oklahoma

Press, Oklahoma.

Dart R. 1948. The Makapansgat proto-human *Australopithecus prometheus Amer. J. Phys. Anthrop.* 6, 259–284.

———— 1959. *Adventures with the Missing Link.* Witwatersrand University Press, Witwatersrand.

Darwin C. 1859. *On the Origin of Species.* John Murray, London.

———— 1871. *The Descent of Man and Selection in Relation to Sex.* John Murray, London.

———— 1872. *The Expression of Emotion in Man and Animals.* John Murray, London.

Davis D. H. S. (Ed), 1964. *Ecological Studies in Southern Africa.* Junk, The Hague.

Day M. H. 1986. *Guide to Fossil Man.* Cassell, London.

Delson E. (Ed), 1985. *Ancestors: the Hard Evidence.* Liss, N.Y.

De Vore E. & Lee R. (Eds), 1976. *Kalahari Hunter-gatherers.* Harvard University Press, Cambridge, Mass.

Diamond J. 1991. *The Rise and Fall of the Third Chimpanzee.* Radius, London.

Dobzhansky T. 1962. *Mankind evolving.* Yale University Press, New Haven.

Doran E. 1981. *Wangka. Austronesian Canoe origins.* Texas & M University Press, Texas.

Drennan M. R. 1931. Paedomorphism in pre-Bushman skull. *Amer. J. Phys. Anthrop.* 16, 203–210.

Durant J. R. (Ed), 1989. *Human Origins.* Clarendon, Oxford.

Dunn L. C. 1959. *Heredity and Evolution in Human Populations.* Harvard University Press, Cambridge, Mass.

Dutrillaux B. 1975. Sur la Nature et l'Origine des chromosomes Humaines. *Monogr. Ann. Genet. Expansion Sci Fr*, 41–71.

Dyer K. F. 1974. *The Biology of Racial Integration.* Scientechnika, Bristol.

Eldredge N. & Tattersall I. 1982. *The Myths of Human Evolution.* Columbia University Press, N.Y.

Erlich A. & Erlich P. 1981. *Extinction.* Random House, N.Y.

Endicott K. 1979. *Batek Negrito Religion.* Oxford University Press, Oxford.

Fagan B. M. 1990. *The Journey from Eden.* Thames & Hudson, London.

Farb P. 1963. *Face of North America. The Natural History of a Continent.* Harper & Bros, N.Y.

———— 1969. *Man's Rise to Civilisation.* Secker and Warburg, London.

Ferembach D. 1962. *La Necropole épipaléolithique de Taforalt.* CNRS, Paris.

Fiedel S. 1987. *Prehistory of the Americas.* Cambridge University Press, Cambridge.

Fisher R. A. 1930. *The Genetical Theory of Natural Selection.* Clarendon, Oxford.

Flood J. 1989. *Archaeology of the Dreamtime,* Collins, Sydney.

Foley R. (Ed), 1984. *Hominid Evolution and Community Ecology.* Academic Press, London.

Forde C. D. 1950–55, *Ethnographic Survey of Africa.* Intern. African Institute, London.

Frayer D. W. 1980. Sexual dimorphism and cultural evolution. *J. Hum. Evol.* 9, 299–415.

———— 1981. Body size, weapon use and natural selection. *Amer. Anthrop.* 83, 57–73.

Freedman L. (Ed), 1989. Is our future limited by our Past? *Proc. 3rd Conf. Austr. Soc. Hum. Biology,* University Western Australia, Perth.

Friedlander J. S. 1975. *Patterns of Human Variation.* Harvard University Press, Cambridge, Mass.

Frisancho A. R. 1979. *Human Adaptation.* Mosby, St Louis.

Fox R. B. 1970. *The Tabon Caves.* Natural History Museum, Manila.

Gamble C. 1986. *The Palaeolithic Settlement of Europe.* Cambridge University Press, Cambridge.

Gamble C. & Soffer O. 1989. *The World at 18,000 B.P.* Unwin Hyman, London.

Garn S. M. 1961. *Human Races.* Springfield, Illinois.

Geerts S. T. 1965. *Genetics Today.* Pergamon, London.

Geist V. 1981. Neanderthal the Hunter. *Natural History* 90 (1), 26–36.

Gillan D. J., Premack D. & Woodruff G. 1981. Reasoning in the chimpanzee. *J. Exper. Psychol. An. Beh. Proc.* 7, 1–17.

Goodall J. 1986. *The Chimpanzees of Gombe.* Harvard University Press, Cambridge, Mass.

Gowlett J. 1984. *Ascent to Civilisation.* Collins, London.

Grine F. E. (Ed), 1990. *The Evolutionary History of the robust Australopithecines.*

Grouchy J., Ebling F. J. & Henderson I. W. (Eds), 1972. Human Genetics. *Excerpta Medica,* Amsterdam.

Groves C. P. 1989. *A Theory of Human and Primate Evolution.* Oxford University Press, Oxford.

Groves C. P. 1989. Natural Selection and Intelligent Ancestors. *Mankind* Vol 19, No 1, 76–82.

Haddon H. C. 1919. *The Wanderings of Peoples.* Cambridge University Press, Cambridge.

Ha Van Tan 1978. The Hoabhinian in Vietnam. *Vietnam Studies* Vol 46, 127–197.

Hamilton A. C. 1982. *Environmental History of East Africa.* Academic Press, London.

Hardin G. 1960. The competitive exclusion principle. *Science* Vol 131, 1291–1297.

Harding R. S. O. & Teleki G. (Eds). 1981. *Omnivorous Primates.* Columbia University Press, N.Y.

Harris D. R. & Hillman G. C. (Eds). 1989. *Foraging and Farming. The Evolution of Plant Exploitation.* Unwin Hyman, London.

Headland T. 1986. *Why Foragers do not become Farmers.* Anne Arbor Univ. microfilm.

———— 1987. The Wild Yam question. *Human Ecology* 15, 463–491.

Headland T. & Reid L. A. 1989. Hunter gatherers and their neighbours from Prehistory to the Present. *Curr. Anthrop.* 30, 43–66.

Heekeren H. R. van. 1972. *The Stone Age in Indonesia.* M. Nizhoff, The Hague.

Hiernaux J. 1974. *The People of Africa.* Weidenfeld & Nicolson, London.

Hill A. *et al.* 1992. Earliest *Homo* in Chemeron Formation at 2.4 MY. *Nature* vol 355, No 6362, 20 February 1992.

Hiorns R. W. & Harrison G. A. 1977. The combined effects of selection and migration in Human Evolution. *Man* 12, 433–445.

Hodges H. 1964. *Artifacts.* Baker, London.

Hope G. S., Golson J. & Allen J. 1983. Palaeoecology and Prehistory in New Guinea. *Jour. Hun. Ecol.* 12, 37–70.

Hornell J. 1934. Indonesian Influence on East African culture. *Journ. Roy. Anthrop. Inst.* Vol. 64, 305–332.

Howell F. C. 1973. *Early Man.* Time Life, N.Y.

Howells W. W. 1966. The Jōmon Population of Japan. A study by discriminant analysis. *Peabody Mus. Papers* 57, 1–43.

———— 1967. *Mankind in the Making.* Addison-Wesley, Reading, Mass.

———— 1973. Cranial Variation in Man. *Papers of Peabody Mus.,* 67, 1–25.

———— 1973. *The Pacific Islanders.* Weidenfeld & Nicholson, London.

———— 1976. Physical variation and history in Melanesia and Australia. *Amer. Journ. Phys. Anthrop.* 45, 641–650.

———— 1976. Explaining modern man; evolutionists versus migrationists. *Journ. Hum. Evol.* 5, 477–495.

———— 1980. *Homo erectus* – Who, When and Where: a Survey. *Yearbook of Phys. Anthrop.* 23, 1–23.

Hughes J. D. 1975. *Ecology in Ancient Civilisations.* University of New Mexico Press, New Mexico.

Hulse F. S. 1971. *The Human Species.* Random House, N.Y.

Huxley J. 1974. *Evolution. The Modern Synthesis.* Allen and Unwin, London.

Huxley T. H. 1863. *Man's Place in Nature.* Macmillan, London.

Iampietro P. F. *et al.* 1959. Response of negro and white males to cold. *Journ. Applied Physiol.* 14, 798–800.

Inizan M. L. & Ortlieb L. 1987. Prehistoire dans las region de Shabwa au Yemen du Sud. *Paleorient* 13, 5–22.

Inskeep R. R. 1978. *The peopling of Southern Africa.* D. Philip, London.

Isaac G. L. 1971. The Diet of Early Man. *World Archaeology* 2, 278–298.

Isaac G. L. & Leakey R. E. F. (Eds), 1979. Human Ancestors. *Scientific American,* Freeman, San Francisco.

Isaac G. L. & McGown E. R. (Eds), 1976. *Human Origins.* Benjamin, Menlo Park.

Jacob T. 1967. *Some problems pertaining to the racial history of the Indonesian Region.* Drukkerie Neerlandia, Utrecht.

Jenkins T., Zoutendyk A. & Steinberg A. G. 1970. Gammaglobulin groups of various S. African populations. *Amer. J. Phys. Anthrop.* 32, 197–218.

Jennings J. D. & Norbeck E. 1956. *Prehistoric Man in the New World.* University of Chicago Press, Chicago.

Jia L. 1980. *Early Man in China.* Science Press, Beijing.

Johanson D. C. & Edey M. A. 1981. *Lucy. The Beginnings of Humankind.* Granada, N.Y.

Johanson D. C. & Shreeve J. 1989. *Lucy's Child.* Morrow, N.Y.

Jolly A. 1972. *The Evolution of Primate Behaviour.* Macmillan, N.Y.

Jolly C. T. 1970. The Seed-eaters: A new Model of Hominid differentiation based upon a Baboon analogy. *Man* Vol 5, 5–26.

Jones J. S. 1981. How Different are Human Races? *Nature* Vol. 293 Sept. 17th, 188–190.

Jones R. (Ed), 1985. Archaeological Research in Kakadu National Park. *Austr. N.P. & WI Service* Publ. 13, Canberra.

Jones R. R. & Wigley T. (Eds), 1989. *Ozone depletion: Health and Environmental Consequences.* Wiley, N.Y.

Kalb J. E., Jolly C. T., Oswald E. B. & Whitehead P. F. 1984. Early Hominid Habitation in Ethiopia. *American Scientist* Vol. 72, 168–178.

Kamminga J. & Wright R. 1988. The Upper Cave at Zhoukoudian and the Origins of the Mongoloids. *Journ. Hum. Evol.* 17, 739–76.

Kano T. 1991. *The last Ape: Pygmy Chimpanzee Behaviour and Ecology.* Stanford University Press, Stanford.

Kennedy J. A. 1984. The Emergence of *Homo sapiens.* The postcranial evidence. *Man* 19, 94–110.

Kimura M. 1983. *The Neutral Theory of Molecular Evolution.* Cambridge University Press, Cambridge.

King J. C. 1981. *The Biology of Race.* University of California Press, Berkeley.

Kinzey W. (Ed), 1985. *Primate Models of Hominid Behaviour*. Plenum, N.Y.

Kirch P. V. 1984. *The Evolution of the Polynesian Chiefdoms*. Cambridge University Press, Cambridge.

Kirk R. L. 1985. *Aboriginal Man adapting*. Clarendon, Oxford.

Kirk R. L. & Thorne A. G. (Eds), 1976. *Origin of the Australians*. Aust. Institute of Aboriginal Studies, Canberra.

Kirk R. L. & Szathmary E. (Eds), 1985. *Out of Asia: Peopling the Americas and the Pacific*. ANU, Canberra.

Klein R. 1969. *Man's Culture in the Late Pleistocene*. Chandler, San Francisco.

————— 1989. *The Human Career*. University of Chicago Press, Chicago.

Kohler W. 1952. *The Mentality of Apes*. Pelican, London.

Kruth G. 1962. *Evolution and Hominisation*. Fischer, Stuttgart.

Lawrence R. 1972. Aboriginal Habitat and Economy. *Austr. Nat. Univ. Dept. Geog. Occ. Paper*, No 6.

Leakey, L. S. B. 1934. *Adam's Ancestors*. London.

————— 1965. *Olduvai Gorge*. Cambridge University Press, Cambridge.

Leakey M. D. 1984. *Disclosing the past*. London.

Leakey R. E. F., Butzer K. W. & Day M. H. 1969. Early *Homo sapiens* remains from the Omo River Region of South West Ethiopia. *Nature* 222, 1132–1138.

Leakey R. E. & Isaac G. (Eds), 1978. *Koobi Fora Research Monograph*. Oxford University Press, Oxford.

Leakey R. E. & Lewin R. 1977. *Origins*. Dutton, N.Y.

Leakey R. E. F. & Walker A. C. 1976. *Australopithecus, Homo erectus* and the Single Species Hypothesis. *Nature* 261, 571–574.

————— 1985. *Homo erectus* unearthed. *National Geographic* 168, 625–629.

Lee R. 1965. *Subsistence Ecology of Kung Bushmen*. PhD Diss. University of California Press, Berkeley.

————— 1979. *Kung San: Men, Women and Work in a Foraging Society*. Cambridge University Press, Cambridge.

Lee R. B. & De Vore I. (Eds). 1968. *Man the Hunter*. Chicago, Aldine.

Le Gros Clark W. E. 1971. *The Antecedents of Man*. Quadrangle, Chicago.

Lewin R. 1987. *Bones of Contention*. Simon & Schuster, N.Y.

Lewontin R. C. 1974. *The genetic Basis of Evolutionary Change*. Columbia University Press, N.Y.

————— 1984. *Human Diversity*. Freeman, San Francisco.

Linden E. 1974. *Apes, Men and Language*. E. P. Dutton, N.Y.

Livingstone F. B. 1962. On the non-existence of Human Races. *Current Anthrop.* 3, 279–282.

Longworth I. H. & Wilson K. E. 1976. *Problems in Economic and Social*

Archaeology. Duckworth, London.

Loomis F. W. 1967. Skin-pigment Regulation of Vitamin D Biosynthesis in Man. *Science* 157, 501–506.

Lovejoy C. O. 1981. The Origins of Man. *Science* 211, 341–350.

———— 1988. Evolution of Human Walking. *Scientific American* Nov. 118–125.

Marshall L. 1960. Kung Bushmen Bands. *Africa* Vol. 30, No 4.

Martin P. & Klein R. 1984. *Quaternary Extinctions*. University of Arizona Press, Tucson.

McNeely J. A. 1988. *Economic and Biological Diversity*. IUCN, Gland.

McBrearty S. 1990. The Origin of Modern Humans. *Man* (NS) 25, 129–143.

Maglio V. J. & Cooke H. B. S. 1978. *Evolution of African Mammals*. Harvard University Press, Cambridge, Mass.

Mann A. 1975. *Palaeodemographic Aspects of the South African Australopithecines*. Philadelphia.

Mascie-Taylor C. G. N. & Lasker G. W. 1988. *Biological Aspects of Human Migration*. Cambridge University Press, Cambridge.

Meehan B. 1977. The Role of Seafood in the Economy of a Contemporary Aboriginal Society in Coastal Arnhemland. *Hansard Rep. 3*, 1085–1095.

———— 1982. *Shell Bed to Shell Midden*. Aust. Institute of Aboriginal Studies, Canberra.

———— 1991. Wetland Hunters: Some reflections. *Monsoonal Australia* 197–206, Balkema, Rotterdam.

Meehan B. & Jones R. (Eds), 1988. *Archaeology with Ethnography*. ANUP, Canberra.

Meehan J. P. 1955. Individual and racial variations in a vascular response to a cold stimulus. *Military Medicine* 116, 330–334.

Meggitt M. J. 1962. *Desert People. A Study of the Walbiri Aborigines*. Angus & Robertson, Sydney.

Mellars P. & Stringer C. (Eds). 1989. *The Human Revolution*. Edinburgh University Press, Edinburgh.

Merimee T. J., Zapf J. & Froesch E. R. 1981. Dwarfism in the Pygmy. *New England Journ. Medic.* 305, 905–908.

Mhoro E. B. & Mtotomwema K. 1988. Mleao Mng'oko: A wild edible Yam. *Tanzania Notes and Records* No 89, 109–126.

Misra V. N. 1987. Presidential Address Proc. 74th Indian Sci. Congress, 1–24, Bangalal, Calcutta.

Misra V. N. & Bellwood P. (Eds), 1985. *Recent Advances in Indo-Pacific Prehistory*. IBH, New Delhi.

Molnar S. 1983. *Human Variation. Races, Types and Ethnic Groups*. Prentice

Hall, N.J.

Montague M. F. (Ed), 1962. *Culture and the Evolution of Man*. Oxford University Press, Oxford.

Montagu A. (Ed), 1964. *The Concept of Race*. Free Press, N.Y.

Moore R. 1960. *Man, Time and Fossils*. Jonathan Cape, London.

Morlan V. J. 1971. Pre-ceramic period in Japan. *Arctic Anthrop*. VIII No 1, 136–170.

Morgan E. 1982. *The Aquatic Ape. A theory of Human Evolution*. Souvenir Press, London.

Mountain M. J. 1979. The Rescue of the Ancestors in Papua New Guinea. *Inst. Archaeol. Bulletin* 16, 63–80.

Mourant A. E. 1983. *Blood Relations: Blood Groups and Anthropology*. Oxford University Press, Oxford.

Mourant A. E., Kopee A. C. & Sobezak K. 1976. *The Distribution of the Human Blood Groups and Other Polymorphisms*. Oxford University Press, Oxford.

Mulvaney D. J. 1969. *The Prehistory of Australia*. Thames & Hudson, London.

Mulvaney D. J. & Golson J. 1971. *Aboriginal Man and the Environment in Australia*. Austr. Nat. University Press, Canberra.

Munroe R. R. 1966. Histological Aspects of Skin Pigmentation. *Arch. Phys. Anthrop. in Oceania* Vol 1 & 2, 119–133.

Napier J. R. 1962. The Evolution of the Hand. *Scientific American* Vol. 206 No 1, 56–62.

Nei M. 1987. *Molecular Evolutionary Genetics*. Columbia University Press, N.Y.

Newman J. L. 1968. *Geography and Subsistence change among the Sandawe*. PhD Diss. University of Minnesota, Minneapolis.

Nishida T. 1990. *The Chimpanzees of the Mahale Mountains*. University of Tokyo Press, Tokyo.

Nitecki M. H. & Nitecki D. (Eds), 1986. *The Evolution of Human Hunting*. Plenum, N.Y.

Nozawa K., Shotake T., Kawamoto Y. & Tanabe Y. 1982. Electrophoretically estimated genetic distance and divergence time between chimpanzee and Man. *Primates* 23, 433–443.

Nurse G. T., Weiner J. S. & Jenkins T. 1985. *The Peoples of Southern Africa and their affinities*. Clarendon Press, Oxford.

Oakley K. P. 1961. *Man the Tool-Maker*. British Museum, London.

Omoto K. 1972. Polymorphism and genetic affinities of the Ainu in Hokkaido. *Human Biology in Oceania* 1, 279–288.

———— 1981. Genetic Origins of the Philippine Negritos. *Current Anthrop*. 22, 421–422.

———— 1987. Population Genetic Studies in the Philippines. *Man and Culture in Oceania* 3, 33–40.

Osborne R. H. (Ed), 1971. *The Biological and Social Meaning of Race.* Freeman, San Francisco.

Pales L. & St Pereuse M. T. 1976. *Les Gravures de la Marché.* Ophrys, Paris.

Passingham R. E. 1982. *The Human Primate.* Freeman, Oxford.

Pearce F. 1990. Hit and Run in Sarawak. *New Scientist* May 12, 24–27.

Perles C. 1975. L'Homme Préhistorique et le Feu. *La Recherche* 60, 829–839.

Perrot A., Street-Perrot A. & Harkness D. 1989. Lake Patzcuaro sediments. *Amer. Antiquities* Vol. 54, 759.

Pfeiffer J. 1978. The Emergence of Man. Harper & Row, N.Y.

———— 1983. *The Creative Explosion.* Harper & Row, N.Y.

Phillipson D. 1985. *African Archaeology.* Cambridge University Press, Cambridge.

Pilbeam D. 1970. *The Evolution of Man.* Thames & Hudson, N.Y.

———— 1972. *The Ascent of Man.* Thames & Hudson, N.Y.

Potts R. 1984. Home Bases of Early Hominids. *American Scientist* 72, 338–347.

———— 1988. On an Early Hominid Scavenging Niche. *Current Anthrop.* 29, 153–155.

Plomley N. J. B. (Ed), 1966. Friendly Mission. *Tasman Hist. Res. Ass.,* Hobart.

Price T. & Brown J. A. 1985. *Prehistoric Hunter-gatherers: The emergence of Cultural Complexity.* Academic Press, N.Y.

Race R. R. & Sanger R. 1964. *Blood Groups in Man.* Thomas Sprinfield, N.Y.

Rak Y. 1983. *The Australopithecine Face.* Harvard University Press, New Haven.

Rambo A. T., Gillogly K. & Hutterer K. L. 1988. *Ethnic Diversity and the Control of Natural Resources in South East Asia.* Michigan University Papers on South East Asia.

Reader J. 1981. *Missing Links.* Little, Brown, Boston.

Reichs K. J. 1983. *Hominid Origins: Inquiries past and present.* University of America Press, Washington.

Renfrew C. (Ed), 1973. *The Explanation of Culture Change.* Duckworth, London.

Roberts D. F. & Hiorns R. W. 1962. The Dynamics of Racial Admixture. *Amer. Journ. Human Genetics* 14, 261–277.

Robinson J. T. 1965. *Homo habilis* and the Australopithecines. *Nature* 205, 121–124.

Roth W. E. 1984, (Facsimile). *The Queensland Aborigines.* Hesperian Press, Perth.

Rowley C. D. 1970. *The Destruction of Aboriginal Society.* ANU Press, Canberra.

Roychoudrhury A. K. & Nei M. 1988. *Human Polymorphic Genes, World Distribution.* Oxford University Press, Oxford.

Russell M. D. 1985. The supraorbital torus: a most remarkable peculiarity. *Current Anthrop.* 26, 336–360.

Saitou N. & Omoto K. 1987. Time and Place of Human Origins from mt DNA. *Nature,* vol. 327, 288.

Salzano F. M. (Ed). 1975. *The Role of Natural Selection in Human Evolution.* Elsevier, N.Y.

Sampson C. G. 1974. *The Stone Age Archaeology of Southern Africa.* Academic Press, N.Y.

Sarich V. M. & Wilson A. C. 1967. Immunological Time Scale for Hominid Evolution. *Science* 158, 1200–1203.

Sawaguchi T. & Kudo H. 1991. Active Selection may participate in the Evolution of Primates. *Human Evolution* Vol 6, No 3, 201–212.

Schapera I. 1951. *The Khoisan Peoples of Africa.* Routledge, London.

Schebesta P. 1928. The Jungle Tribes of the Malay Peninsula. *Bull. of School of Orient. and Afr. Studies* 4, 269–278.

Schrire C. (Ed) 1985. *Past and present in Hunter-gatherer studies.* Academic Press, N.Y.

Service E. R. 1966. *The Hunters.* Yale University Press, N.J.

Sheldon W. H. & Tucker W. B. 1941. *The varieties of Human Physique.* Hafner, N.Y.

Shepard C. 1831. *An Historical Account of the Island of St Vincent.* Weidenfeld & Nicolson, London.

Sherratt A. (Ed), 1980. *The Cambridge Encyclopaedia of Archaeology.* Cambridge University Press, Cambridge.

Shipman P. 1984. Scavenger Hunt. *Natural History,* April, 20–27.

———— 1984. The earliest Bone Tools. *Anthroquest* 29, 9–10.

Shipman P., Bosler W. & Davis K. L. 1981. Butchering of Giant Geladas at an Acheulian site. *Current Anthrop.* 22, 257–264.

Shipman P. & Walker A. 1989. The Costs of becoming a Predator. *J. Human Evolution* 18, 373–392.

Sibley C. G. & Alquist J. E. 1984. The Phylogeny of the Hominid Primates as indicated by DNA–DNA Hybridization. *Journ. of Molecular Evolution* 20, 2–15.

Simmons R. T., Tindale N. B. & Birdsell J. B. 1962. A Blood Group genetical Survey in Australian Aborigines. *Amer. Journ. Physic. Anthrop.* Vol. 20, No 3, 303–320.

Simpson, G. G. 1967. *The Meaning of Evolution.* Yale University Press, New Haven.

Singh R. 1975. Arrows speak louder than words. The last Andaman Islanders. *National Geographic,* Vol. 148, July 1975, 66–76.

Smith F. & Spencer F. 1984. *The origins of Modern Humans.* A. Liss, N.Y.

Soffer O. (Ed). 1987. *The Pleistocene Old World.* Plenum, N.Y.

Standen V. & Foley R. A. 1991 (Eds). *Comparative Socioecology of Mammals and Man.* Blackwells, Oxford.

Stern C. 1973. *Principles of Human Genetics.* Freeman, San Francisco.

Stocking G. W. (Ed). 1973. *Researches into the Physical History of Man.* University of Chicago Press, Chicago.

Stringer C. 1974. Population Relationships of later Pleistocene Hominids. *J. Archaeol. Sci. I,* 317–342.

———— (Ed). 1981. *Aspects of Human Evolution.* Taylor and Francis, London.

Stringer C. & Andrews P. 1988. Genetic and Fossil Evidence for the origin of Modern Humans. *Science* 239, 1263–1268.

Susman R. L. (Ed). 1984. *The Pygmy Chimpanzee.* Plenum, N.Y.

Symons D. 1979. *The Evolution of Human Sexuality.* Oxford University Press, Oxford.

Szalay F. S. (Ed), 1975. *Approaches to Primate Palaeobiology.* Karger, Basel.

Tattersall I. 1986. Species Recognition in Human palaeontology. *Journ. Hum. Evol.* 15, 165–175.

Terrell J. 1986. *Prehistory in the Pacific Islands.* Cambridge University Press, Cambridge.

Thomson M. L. 1955. Relative Efficiency of Pigment and Horny Layer thickness in protecting the skin of Europeans and Africans against solar Ultraviolet Radiation. *J. Physiol.* 127, 236–246.

Tiger L. & Fowler H. (Eds). 1978. *Female Hierarchies.* Beresford B.S., Chicago.

Tindale N. B. & Lindsay H. A. 1963. *Aboriginal Australians.* Angus and Robertson, Sydney.

Tobias P. V. 1955. Physical Anthropology and the Somatic origins of the Hottentots. *African Studies* 14, 1–15.

———— 1965. Early Man in East Africa. *Science* 199, 22–33.

———— 1971. *The Brain in Hominid Evolution.* Columbia University Press, N.Y.

———— (Ed). 1985. *Hominid Evolution Past, Present and Future.* A. Liss, N.Y.

Tomkins S. 1984. *The Origins of Mankind.* Cambridge University Press, Cambridge.

Trinkaus E. (Ed). 1991. *Corridors, Cul-de-Sacs and Coalescence. The Biocultural Foundations of Modern People.* Cambridge University Press, Cambridge.

Turnbull C. 1961. *The Forest People.* Pan, London.

Turner C. G. 1987. Late Pleistocene and Holocene Population History of East Africa based on Dental Variation. *Amer. Journ. Phys. Anthrop.* 73, 305–321.

Tuttle R. H. (Ed). 1976. *Primate Morphology and Evolution.* Junk, The Hague.

Van Valen L. M. 1986. Speciation and our own Species. *Nature* 322, 412.

Walker A. C. & Leakey R. E. 1978. The Hominids of East Turkana. *Scientific American* 239, (8), 54–66.

Walker D 1972. *Bridge and Barrier, the Natural and Cultural History of the Torres Straits.* Junk, The Hague.

Washburn S. L. (Ed). 1963. *Social Life of Early Man.* Viking, N.Y.

————— (Ed). 1964. *Classification and Human Evolution.* Aldine, Chicago.

————— 1968. *The Study of Human Evolution.* Oregon.

Washburn S. L. & Dolhinow P. (Eds), 1972. *Perspectives on Human Evolution.* Holt Rinehart, N.Y.

Wasson R. J. & Clark R. 1987. Sea Levels at Huon peninsula. BMR Report, 282.

Watanabi H. 1985. The chopper-chopping tool complex of Eastern Asia. *Journ. Anthrop. Archaeol.* 4, 1–18.

Watanabe S. (Ed). 1975. *Anthropological and Genetic Studies of the Ainu.* Tokyo.

Weiner J. S. 1954. Nose-shape and Climate. *Amer. Journ. Phys. Anthrop.* 12, 1–4.

Weiss K. M. 1984. On the number of Members of the Genus *Homo* who have ever lived and some evolutionary Implications. *Human Biology* 56, 637–649.

White T. D. 1980. Evolutionary Implications of Pliocene Hominoid Footprints. *Science* Ap 208, 175–176.

White P. & O'Connell J. (Eds). 1982. *A Prehistory of Australia, New Guinea and Sahul.* Academic Press, Sydney.

Wilson E. O. 1984. *Biophilia.* Harvard University Press, Cambridge, Mass.

————— (Ed). 1988. *Biodiversity.* Nat. Academic press, Washington.

Wolpoff M. H. 1968. Climatic Influence on the skeletal nasal Aperture. *Amer. Journ. Phys. Anthrop.* 29, 405–423.

Wood B. A. 1978. *Human Evolution.* Chapman and Hall, London.

————— 1987. Who is the 'real' *Homo habilis? Nature* 327, 187–188.

Wood B., Martin L. & Andrews P. (Eds). 1984. *Major Topics in Primate and Human Evolution.* Cambridge University Press, Cambridge.

Wu R. & Lin S. 1985. *Chinese Palaeoanthropology: Retrospect and Prospect.* Beijing.

Wu R. & Olsen J. W. (Eds). 1985. *Palaeoanthropology and Palaeolithic Archaeology in the People's Republic of China.* Academic Press, N.Y.

Yellen J. E. 1977. *Archaeological Approaches to the Present; Models for reconstructing the Past.* Academic Press, N.Y.

Yoffee N. & Cowgill G. L. 1988. *The Collapse of Ancient States and Civilisations.* University of Arizona Press, Tucson.

Young J. Z. 1974. *An Introduction to the Study of Man.* Oxford University Press, Oxford.

Young J. Z., Jope E. M. & Oakley K. P. (Eds), 1981. The Emergence of Man. *Phil. Trans. Roy. Soc.* 292.

Young V. 1795. *An Account of the Black Caribs in the Island of St Vincent.* J. Sewell, London.

Yunis J. J. & Prakash O. 1982. The Origin of Man: a chromosomal pictorial Legacy. *Science* 215, 1525–1530.

Zihlman A. L., Cronin J. E., Cramer D. L. & Sarich U. M. 1978. Pygmy Chimpanzees as a possible Prototype for the common Ancestor of Humans, Chimpanzees and Gorillas. *Nature* 275, 1525–1530.

Zuckerman S. 1966. Myths and Methods in Anatomy. *Journ. Roy. Coll. of Surgeons, Edinburgh,* 87–114.

Index